Instructor's Manual

to accompany

Understanding

Fiber Optics

Fifth Edition

Jeff Hecht

While every precaution has been taken in the preparation of this book, the publisher assumes no responsibility for errors or omissions, or for damages resulting from the use of the information contained herein.

INSTRUCTOR'S GUIDE TO ACCOMPANY UNDERSTANDING FIBER OPTICS FIFTH EDITION

First edition. August 11, 2022.

ISBN: 979-8223300830

Written by Jeff Hecht.

Laser Light Pres

Preface:

About *Understanding Fiber Optics, 5th ed*

Understanding Fiber Optics began as a self-study text, but has been widely adopted for classroom use. It concentrates on fiber optics for communications, but also gives brief introductions to other uses of fiber optics in imaging, illumination, and sensing. It is written for the classroom rather than as a laboratory text; my goal is to explain principles rather than to detail procedures such as cable installation. Several chapters, including the first three introductory chapters, have been rewritten and reorganized. The entire book also has been updated to clarify concepts and to reflect new technology and business developments such as the start of deployment of fiber to the home systems. Many illustrations have been redrawn to make them clearer and to correct mistakes. The changes include more realistic descriptions of the actual use of fiber systems, and discussions of how the telecommunications bubble distorted reality. To help you keep up to date, I have included lists of written and organizational resources at the end of this manual, as well as notes on Web resources.

My goal is to explain fundamental concepts at an elementary level accessible to anyone with a minimal technical background. To that end, I have written in a casual tone, included many "cartoon" drawings showing how things work, taken pains to explain industry jargon before using it, and generally avoided math beyond basic algebra. I draw analogies and contrasts with common technologies, particularly electronics, both to help students understand how fiber optics work, and to illuminate important differences. I assume students taking a fiber optics course have had some exposure to electronics, but I don't assume an in-depth expertise in either electronics or communications.

I use metric units throughout, because they are standard for most dimensions in optics and fiber optics. I translate metric units into Imperial (English) units in only a few places. The communications industry is inconsistent in its usage; while fiber attenuation is universally measured in decibels per kilometer, cable runs are often given in miles or feet.

Each chapter includes a multiple-choice quiz for students to test themselves, and a set of open-ended "Questions to Think About" to push their horizons. The questions span a broad range, from asking "why" something works (or doesn't work) to word problems requiring calculations. The fifth edition for the first time includes short "things to think about" in many chapters, covering ideas from whether glass can flow at ordinary temperatures to digital rights management on the Internet.

This instructor's manual is intended to give you a quick overview of what each chapter contains, and to suggest teaching strategies and additional resources. The description of each chapter includes answers to the multiple choice quiz and to the open ended "Questions to Think About" included with each chapter in the text; to save you the trouble of looking back and forth, I repeat the Questions to Think About included in the text.

The CD-ROM that accompanies this instructor's manual includes art in PowerPoint form for your classroom use, based on short courses I have taught for the Optical Society of America and SPIE, plus copies of illustrations from the book.

This edition is organized into several blocks of chapters.

The first three chapters are an introductory overview. Chapter 1 is a historical introduction to the field and to some fundamental concepts used throughout the book. Chapter 2 introduces basic concepts of optics and physics, then shows how they explain the basic principles of fiber optics and photonics. Chapter 3 introduces communications systems and concepts in general terms that cover both electronic and optical communications. Inevitably students have uneven preparation in basic optics and communications, and these chapters are designed to smooth over any holes in their backgrounds to prepare them for later chapters. How much class time you devote to these chapters depends on your students' preparations, but you should assign them as background reading.

Chapters 4 through 8 cover optical fibers and cables. To keep the information in manageable chunks, I divided the coverage of fibers into separate chapters on fiber types, properties, materials, and specialty fibers. Chapters 4 through 6 are essential to understanding the fiber concepts spread through the rest of the book. Chapter 7 covers fibers developed for purposes other than transmitting signals between points—including fiber gratings, reduced-core fibers, polarization-maintaining fibers, and the rare-earth-doped fibers used in optical amplifiers. Chapter 8 explains the structure and function of cabling.

Chapters 9-12 cover optical sources, amplifiers, and receivers. Light sources are treated as components in Chapter 9, which covers their operating principles. Chapter 10 covers transmitters, including multiplexing and modulation as well as light sources. Chapter 11 covers both optical detectors and receivers. Chapter 12 covers repeaters, regenerators and optical amplifiers. These chapters give few circuit examples, but don't go into much detail because this is a book on optics rather than electronics.

Chapters 13-16 cover various additional "nuts and bolts" of fiber-optic systems. Chapter 13 integrates connectors and splices to highlight their differences and similarities. Chapter 14 covers couplers and other passive components such as optical attenuators, optical isolators, and optical circulators. Chapter 15 is devoted to optics used for wavelength-division multiplexing. Chapter 16 covers active components including modulators, optical switches, and wavelength converters.

Test and measurements are the subjects of Chapters 17 and 18. Chapter 17 introduces the principles of optical measurements and important techniques. Chapter 18 covers test equipment and troubleshooting.

Note that Chapters 4-18 concentrate on *optical* components and measurements. They look at the building blocks of fiber-optic systems, rather than at the overall operation of communication systems. They also concentrate on the optical aspects of these devices, which are essential to understanding fiber optics. This grouping provides a logical break to review optical ideas and to run a question-and-answer session to help students fill gaps in their understanding. If you plan to cover *only* the component side of fiber optics, you may skip most of Chapters 19-28 which deal with communication systems, and include Chapters 29 and 30, which cover non-communication uses of fibers. If systems are addressed in a different course which students have not yet taken, the material in Chapter 19-28 may be helpful as examples of how components are used.

Chapters 19-28 cover the principles and operation of fiber-optic systems and optical networking. These chapters are not intended as a comprehensive course in communications, but they do introduce important communication concepts needed to understand the applications of fiber optics in communications. Thus these chapters concentrate more on the transmission network where fiber-optics are used than on parts of the network where fibers have little role, such as home telephone instruments. The treatment assumes that students know little about communications systems, so students who have already had communications technology courses can skim the background material. The emphasis is on technology rather than on industry structure or regulations.

Chapter 19 introduces system and optical networking concepts, with particular emphasis on transmission. Chapter 20 covers the standards that impact fiber-optic networks. Chapter 21 explains the design of fiber-optic systems carrying only a single optical channel. Chapter 22 explains concepts of optical network design and wavelength-division multiplexed systems. Chapter 23 describes long-distance transmission on land and in submarine cables. Chapter 24 describes regional and metro telecommunications. Chapter 25 covers local telephone or "access" networks and fiber to the home or premises systems. Chapter 26 covers data transmission, local area networks, and the Internet. Chapter 27 covers video transmission with emphasis on cable television. Chapter 28 covers the use of fiber optics for remote control of robotic vehicles, and within vehicles ranging from battleships to automobiles.

The final two chapters can be skipped if your course covers only fiber-optic communications. However, sensing (Chapter 29) and image transmission and illumination (Chapter 30) are important applications for fiber optics.

Appendices A through D tabulate information that students may find useful during and after the course. Appendix A compiles important formulas and conversion factors mentioned in the text; they are gathered in one place to make life easier. Other appendices include tables of decibel equivalents, standard data rates, and standard optical channels. Appendix E gives a very basic introduction to laser and fiber safety. Appendix F lists resources valuable for further research.

The glossary defines important fiber-optic terms and acronyms used in the text. Telecommunications is plagued by a bewildering variety of acronyms, which I try to avoid. One reason is philosophical—acronyms are hard to assimilate, so I find it counterproductive to throw them at a beginning student entering a new field. Another reason is pragmatic—few acronyms are standardized. Some appear only in specific articles or books; others are used only by a few technical journals, industry magazines, or corporate publications. The back of this instructor's guide includes a list of some extra acronyms that I don't use or that didn't appear in the glossary, but be aware that it is not comprehensive.

As a frequent user of reference books, I believe a good index is a vital part of textbooks like this, and have compiled one to make *Understanding Fiber Optics* valuable as a reference book during and after the course. Please urge your students to check the index if they have questions.

My home page for fiber optics is **http://www.fiberhome.com**. At this writing it is devoted to a brief description of this book, but I hope to add further material and links. You can find information on other books I have written at **http://www.jeffhecht.com/books/html**.

Although I do teach some short courses, my primary activity has long been writing about technology. You can find my articles in magazines including *Laser Focus World*, *New Scientist*, and *Optics & Photonics News*. I particularly recommend the articles in *Laser Focus World* and its companion magazine *Lightwave* as a way to update yourself on fiber optics.

The sections that follow briefly outline the goals of each chapter, give answers to questions, and suggest supplementary material. They explain answers to the open-ended "Questions to Think About", and include some worked-out numerical solutions to selected multiple-choice problems.

I welcome any suggestions and comments you might have for future editions. If you think you have found mistakes, or if anything seems unclear, please contact me. I am far from infallible, and continue searching for ways to improve my explanations. Please contact me by mail through Prentice-Hall, or e-mail me at jeff@fiberhome.com or jeff@jeffhecht.com.

—Jeff Hecht, Auburndale, MA, March 2005

Chapter 1

Introduction to Fiber Optics

Chapter 1 is an introduction and overview, to give students a general background in fiber optics and its uses. Depending on the preparation of your students, you could use this material in an introductory lecture, or assign the chapter as outside reading.

Overview

The development of any technology can be a helpful blueprint to understanding its workings, so I have organized this explanation in historical sequence. Fiber optics began with the idea of using total internal reflection to guide light in jets of water, first demonstrated in lectures and later at the great Victorian exhibitions of the late 1800s.Later engineers expanded the idea of light guiding to groups of fibers bundled together to carry images, with the major applications in medicine and military imaging systems. Communications came only later, and required major improvements in glass transparency. (Note that many brief histories of early fiber optics include significant mistakes, starting with attributing the water jet experiment shown on page 4 of the book to John Tyndall.)

I find the history fascinating, but the real importance for the beginning student is to introduce the central concepts of total internal reflection, light guiding, the fiber cladding, bundles, imaging, and communications. It's also important for students to learn that ideas do not spring full-blown from the mind of a single inventor, but instead evolve over the years, as new ideas emerge in response to new technologies, capabilities, and new potential applications.

After describing the history, the chapter moves on to basic fiber-optic concepts. Later chapters cover these ideas in more detail, but my approach is to start with simple explanations for the least-prepared students. From there, the chapter moves to an introduction to the role of fiber optics in communications.

I also describe the business of telecommunications and the disruption caused by the Internet and the bubble. I think the booms and busts of the 19th century railroad industry are a good analogy for what happened to fiber optics. Like the railroads revolutionized ground transportation, fiber optics revolutionized telecommunications. Both technologies succeeded so well that investors poured tremendous sums of money into them, and their capacity was expanded far beyond any immediate needs, causing a bust. Essentially, the industry got running so fast that it ran right off a cliff, and like Wyle E. Coyote, didn't realize what had happened until it looked down.

Questions to Think About For Chapter 1

1. FOR A BUNDLE OF optical fibers to transmit an image, the fibers must be arranged in the same pattern on both ends of the bundle. What limits the size of the smallest details that can be seen?

Answer: The diameter of the fiber, because all light entering the fiber is mixed as it travels along the fiber.

2. Devise an analogy to show how a bundle of fibers transmits an image using common implements found in a kitchen or cafeteria.

Answer: A bundle of drinking straws stacked parallel to each other.

3. Most of the light lost in going through a glass window is reflected at the surface. Ignoring this surface reflection loss, suppose that a one-millimeter window absorbs 1% of the light entering it and transmits 99%. Neglecting reflection, how much light would emerge from a one-meter thick window?

Answer: 99% passes through each millimeter, so you multiply 0.99 times itself 1000 times, calculating 0.99^{1000}. The answer is 0.000043, or 43×10^{-6} of the light entering the glass. For comparison, that's about the difference between the magnitude of the planet Venus at its brightest and the faintest star visible to the unaided eye on a dark night far from city lights.

4. If optical fibers transmit signals so much better than wires, why aren't they used everywhere?

Answer: Largely because telephone companies and others have installed billions of dollars worth of existing copper cables. In addition, input signals come in electronic form, and must be converted into optical form, which requires separate optical transmitters and receivers.

5. During the bubble years, many people in the industry thought Internet traffic was doubling every three months. In reality, it was doubling about every year. How much difference would this make over a period of five years?

Answer: If traffic doubled every year, it would be 32 times higher after 5 years. If traffic doubled every 3 months, it would be more than one million times higher (there are 20 three-month periods in five years, and $2^{20}=1,048,576$). The estimate that traffic was doubling every month came from WorldCom, which time has proved was not the most reliable source.

6. Why didn't anybody wonder how long Internet traffic could continue doubling every three months?

Answer: A few people did, and it was Andrew Odlyzko, now at the University of Minnesota, who first showed data to contradict those claims, but most people either were carried away by the mania of the bubble, or didn't have data to question it.

ANSWERS TO QUIZ FOR Chapter 1

1. c

2. d

3. a

4. d

5. b

6. a

7. d

8. d

9. c

10. a

<u>Supplemental Materials and Suggestions:</u>

A sample of fiber (preferably flexible thick plastic fiber for safe and easy handling) to pass around the classroom. Fiber-optic Christmas ornaments are cheap and should be readily available. Handling fibers is a good way to understand how they work, and plastic fibers are much safer than glass to pass around.

The water-jet experiment is not as simple as it looks, so don't expect to recreate it easily.

For more on the history of fiber optics and the bubble, see Jeff Hecht, *City of Light: The Story of Fiber Optics* (Oxford University Press, 1999, 2nd edition 2004).

A good example of how bad the collapse of the bubble hit telecommunication stocks is a comparison of a couple of rather different investments. If you'd invested $1000 in Nortel stock about September 2000, you would have had $72 a year later. If you'd invested $1000 at the same time in Budweiser—the beer, not the stock—and lived in a state with a bottle deposit law, you would have had $76 worth of empties.

Chapter 2

Fundamentals of Fiber-Optic Components

This chapter explains the basic principles of optics, optical fibers, and fiber-optic components that are essential to understanding later chapters. This is the starting point for your "meaty" lectures. You should open with the basics of refraction, and show how those principles lead to total internal reflection. Then you should cover light guiding in fibers (using the total internal reflection model), and other properties of fibers and optical components.

<u>Overview</u>

The starting point is basic optics, because students must understand light and its properties before they consider fiber optics. Much of this information may not be new to students who have had a good introductory physics course, but you can't count on that. This review focuses upon the parts of optics relevant for fiber optics.

I start with the ideas of the electromagnetic spectrum, wavelength, frequency, and photons. Then I introduce refraction and refractive index, which leads to total internal reflection, and how it can explain light guiding in a multimode step-index optical fiber. That explanation of light guiding takes the traditional optical perspective of tracing the paths of light rays, rather than the more accurate treatment of an optical fiber as a waveguide, because the ray model is more intuitive for students. The waveguide model is introduced in Chapter 4, which also introduces the graded-index and single-mode fibers.

This chapter also introduces fiber properties, including attenuation coupling losses, pulse dispersion and transmission bandwidth. These are fundamental concepts that appear throughout the book, and are described in more detail in later chapters. They are introduced here to give students a working background for later chapters. You can refer students who want more details to the index. A brief final section introduces optical transmitters, receivers, amplifiers, and repeaters.

Students should finish this chapter with a general knowledge of fiber optic components, adequate to explain the subject to the proverbial man on the street, and to comprehend more detailed descriptions.

<u>Questions to Think About for Chapter 2</u>

1. Interference seems to be a strange effect. The total light intensity from two bulbs is the sum of the two intensities. Yet the light intensity is really the square of the amplitudes, and if the two waves are in phase, you double the amplitude, which when squared means the intensity should be four times the intensity of one bulb. Don't these views contradict each other?

Answer: Not really because interference varies from place to place. If the bulbs emitted identical light waves, they would interfere constructively in some spots but destructively in others. Averaged over space, they're halfway between completely in phase (where intensity would be four times that of one bulb) and completely out of phase (where intensity would be zero), so intensity would average to that of two bulbs. If you scanned over the whole volume with a detector small enough to see the variations on the scale of a wavelength, you would see intensity varying continually between total darkness and peak light.

2. One photon is a wave packet that doesn't last very long. A continuous light source emits a steady or continuous wave. How is the continuous light source emitting photons?

Answer: In a steady stream. If the light source emits a steady amount of energy per second, it emits a constant number of photons each second.

3. The sun emits an energy of about 3.8×10^{33} ergs per second. A photon with wavelength of 1.3 micrometers has an energy of about 1.6×10^{-12} erg. If you assume the sun emits all its energy at 1.3 μm, How much attenuation in decibels do you need to reduce the sun's entire output to a single 1.3-μm photon per second?

Answer: The ratio is 2.375×10^{45}, which corresponds to 454 dB.

4. If an entire galaxy contains a billion stars, each one as luminous as the sun, how much attenuation does it take to reduce its entire output to a single 1.3-μm photon per second.

Answer: The ratio is 2.375×10^{54} or 544 dB.

5. Suppose window glass has attenuation of 10 dB/meter at 1.3 micrometers (not an actual value). How thick a block of glass would you need to reduce the sun's entire output to a single photon as in problem 3?

Answer: Divide 454 dB by 10 dB/m to get 45.4 meters.

6. Medical imaging fiber has attenuation of 1 dB/meter at optical wavelengths. If the attenuation is the same at 1.3 μm, and you don't have to worry about the sun's energy melting the fiber, how long a fiber would reduce the sun's output in problem 3?

Answer: Divide 454 dB by 1 dB/m, giving 454 meters.

7. Atoms and molecules in the atmosphere scatter light in the same way that atoms in glass scatter light in an optical fiber. The shorter the wavelength in the visible spectrum, the stronger the scattering. Where do you think the sky gets its blue color from and why?

Answer: Blue and violet light have the shortest wavelengths in the visible spectrum, but the eye isn't very sensitive to violet. The blue light is scattered much more than longer wavelengths. The blue light you see is scattered sunlight.

8. Diamond has a refractive index of 2.4. What is its critical angle in air and what does that have to do with its sparkle.

Answer: The critical angle is 28°, so it reflects light by total internal reflection over a wider range of angles than glass. Diamonds are cut so the facets at the back of the stone will reflect light back to your eye.

<u>Worked Quiz Problems</u>

Problem 7:

The problem implicitly assumes the zircon is in air. The critical angle is

$$\theta_{crit} = \arcsin\left(\frac{n_{air}}{n_{zircon}}\right) = \arcsin\left(\frac{1}{2.1}\right) = 31.6°$$

PROBLEM 9:

First calculate total loss in decibels:

$$Loss = 0.3db/km \times 100km = 30dB$$

Then convert that loss back to a percentage:

$$Ratio = 10^{\left(\frac{loss\,in\,dB}{-10}\right)} = 10^{\left(\frac{30}{-10}\right)} = 10^{-3} = 0.1\%$$

<u>Answers to Quiz for Chapter 2</u>

1. e

2. e

3. c

4. b

5. c

6. d

7. c

8. c

9. a

10. d

11. a

12. b

<u>Supplemental Material and Suggestions</u>

Illustrations showing light refraction and the operation of lenses, if available.

Shine a laser pointer through a cloud of chalk dust to show how light travels in straight lines. (Cigar smoke also works well.)

Chart of the optical spectrum to illustrate wavelengths and colors. Point out that most fiber-optic transmission is at 850, 1300 and 1550 nm.

Plot of fiber loss to show fiber transmission windows; see Figure. 5.2.

David Falk, Dieter Brill, and David Stork's *Seeing the Light: Optics in Nature, Photography, Color, Vision, and Holography* (Wiley, 1986) is a wonderful nonmathematical treatment of optics.

J. Warren Blaker and Peter Schaeffer, *Optics: An Introduction for Technicians and Technologists* (Prentice Hall, 2000) is a short primer on optics for technicians, with some math, which goes into more detail on optical components other than fibers.

Chapter 3

Fundamentals of Communications

This chapter is an introductory overview of communications systems and services, intended to provide the background needed to understand fiber-optic components and systems and how they are used. It covers basic concepts and applications, then turns to fiber optics. Although software is important in communications, my emphasis is on hardware because fiber optics fall in that category. If your students have had an introductory communications course, this chapter can serve as a review, and you may not need to go through it in class. The chapter introduces terms used in the rest of the text, which can be important because communications terminology is far from universal.

Overview

The first part of the chapter introduces the concept of telecommunications, and describes the historical evolution of communications technology. This puts fiber optics into the context of the growing need for bandwidth. I also introduce important terms the student is likely to encounter, giving fuller descriptions than is possible in a glossary, and revealing my prejudices about the most meaningless buzzwords. This section also introduces the concept of multiplexing, and defines frequency-division, wavelength-division, and time-division multiplexing. Note that I use the term "optical channel" to mean a signal carried by one wavelength in any kind of optical system, but it generally applies to wavelength-division multiplexing.

The next section covers signal formats, explaining carriers and modulation. I give examples of carriers and modulation. The section on analog and digital communications is important in explaining the difference, and in showing how an analog signal can be converted to digital form. It also introduces the difference between optical signals, as a number of photons, and electronic signals, as a voltage or current.

The next section describes the types of connectivity offered by fiber-optic systems, divided loosely into broadcast systems, switched systems, and networked systems. It also introduces switching technologies.

The following section covers communication services in general, emphasizing the telephone network, cable television, and the Internet. It also introduces the concept of convergence, which will be important in the systems section of the book.

The final section outlines the business side of telecommunications, listing a number of different types of carriers and other elements of the industry, and introduces the importance of regulations in telecommunications. The treatment throughout is elementary, but the concepts are critical to further understanding.

Questions to Think About for Chapter 3

1. How does using a higher-frequency carrier affect the amount of information that can be transmitted?

Answer: It increases bandwidth.

2. Why does multiplex transmission of a combined signal cost less than separate transmission of each signal?

Answer: Each multiplex transmitter generally costs more than a single-channel transmitter, but you need fewer of them. Likewise, the transmission medium often costs more, but the cost is not multiplied by as large a number as the capacity.

3. Computer networks, mobile telephones and broadcast systems all distribute signals to many terminals. How do these systems differ?

Answer: Computer terminals and mobile telephones ignore signals that are not addressed to them. Broadcast signals are picked up and decoded by all terminals within range.

4. Why is frequency-division multiplexing equivalent to wavelength-division multiplexing?

Answer: Wavelength and frequency are two different ways of measuring the same property of an electromagnetic wave. Radio engineers usually speak of frequency, while optical engineers usually talk of wavelength. Interestingly, the standard spacing for wavelength-division multiplexing channels is specified in frequency units.

5. The bandwidth of digitized signals measured in bits per second is much higher than the bandwidth of the original analog signal measured in hertz. For example, the analog bandwidth of a phone line is 4000 Hz, but the digitized signal is 64,000 bits per second. Why are these two bandwidths usually equivalent in practice?

Answer: Analog signals must be reproduced much more accurately than digital signals, so an analog transmission line has to carry some higher-frequency harmonics. A digital system doesn't need to reproduce frequencies higher than the bit rate.

6. Why was telephone service considered a natural monopoly?

Answer: It was considered impractical to build multiple telephone networks to serve the same community.

7. Data transmission rates to personal computers have increased from 1200 bits per second with dial-up modems in 1985 to about 400,000 bits per second with a cable modem or DSL in 2000. If bandwidth keeps increasing at the present rate, how fast will transmission be in 2015?

Answer: A linear extrapolation is 133 megabits per second.

<u>ANSWERS TO QUIZ FOR Chapter 3</u>

1. b

2. e

3. e

4. d

5. a

6. d

7. c

8. a

9. a

10. c

11. d

12. d

<u>Supplemental Material</u>

Illustrations or maps of communication systems. A good source is **http://www.cybergeography.org**

Examples of voice, video, and data communications.

Demonstrate analog-to-digital and digital-to-analog conversion. A blow-up showing the discrete levels in a digital signal would help.

Mention any fiber networks in your facility or nearby that students might know.

Ask students what uses of fiber optics they have seen, both in communications and in other fields. (Note that some endoscopes uses fibers to deliver light which is picked up by a tiny video camera on the end of the device inside the body.)

Chapter 4:

Types of Optical Fibers

This chapter describes how different types of fibers work and how they are used, introducing transmission modes. These topics are among the most important topics for you to cover in lectures. Types of fibers specified by ITU standards are now identified by the standard number as well as by their common descriptions.

<u>Overview</u>

This chapter covers major fiber types largely in the historical order of their development. This largely follows the level of increasing conceptual complexity. I start by describing light guiding in step-index multimode fibers using the traditional model of total internal reflection of optical rays. Then I introduce waveguides and modes, and explain how that model is a more complete description of light guiding than the simpler view of total internal reflection.

The description of modes is targeted at the introductory level, and includes only algebraic equations. I avoid derivations except the few that can be done clearly with simple algebra, an approach I take throughout the book with very few exceptions. This approach gives students what they need to know, without discouraging them with complex equations. I introduce modes to describe single-mode fibers, and move from that to graded-index fibers.

The balance of the chapter covers various types of single-mode fibers, starting with the simplest case, standard step-index single-mode fiber. Then I introduce the concept of chromatic dispersion, and describe the concept of dispersion shifting. That leads into descriptions of the different types of dispersion-shifted fibers. I show a sampling of refractive-index profiles, but do not explain the design of dispersion-shifted fibers, which requires too much math for this level. The final brief section explains polarization in single-mode fibers.

The dispersion problem has shaped fiber design in many ways, so I have interwoven descriptions of dispersion with those of fiber types. The importance of modal dispersion was recognized very early, and the first proposals for fiber-optic communications envisioned using single-mode fiber. Step-index multimode fiber was never considered useful for communications over any significant distance because of its high modal dispersion. Graded-index multimode fiber was developed to reduce modal dispersion while retaining the advantage of large core diameter; it was used in a few early telephone systems, but now is used mostly in local-area and campus networks spanning no more than a few kilometers. New graded-index fibers allow much higher bandwidth than previously, but remain limited to relatively short distances.

Single-mode fiber is used in most telecommunications networks, including fiber to the premises or home. The most widely used design is "standard" step-index single-mode fiber, specified by the ITU G.652 standard. Dispersion-shifted fibers normally are used only in long systems where dispersion management is required.

<u>Questions to Think About for Chapter 4</u>

1. A step-index multimode fiber has modal dispersion of about 30 ns/km. Using the formula for maximum data rate for a given dispersion, about how far could it transmit a signal at 1 Gbit/s?

Answer: The formula specifies Data Rate =0.7/(dispersion), a formula for No-Return-to Zero signals. Dispersion in this formula is the characteristic dispersion (per unit length) times the length. Thus

$$Data\,rate = \frac{0.7}{(30\,ns/km) \times Length(km)}$$

Inverted to calculate the maximum transmission length, this becomes

$$Length(km) = \frac{0.7}{(30 \times 10^{-9}\,s/km) \times 10^{9}\,bits/s} = 0.023$$

Or 23 meters. Thus step-index multimode fiber is like twisted pair copper wires—it can carry high-speed signals, but not very far.

2. Why doesn't dispersion affect imaging or illumination fibers?

Answer: It doesn't matter because the transmitted signals are not varying with time.

3. Graded-index fiber typically is more expensive than step-index single-mode fiber. Yet it is used to carry Gigabit Ethernet signals several hundred meters. What advantage does it offer?

Answer: The larger core diameter makes it easier to get light into graded-index fiber, reducing the cost of transmitters and connectors.

4. What are the tradeoffs between effective area and dispersion slope?

Answer: Normally reduced dispersion slope fibers have small effective areas. Reducing dispersion slope reduces the variation of dispersion with wavelength, and makes dispersion easier to compensate for. However, small effective area fibers are more vulnerable to crosstalk because the light is concentrated in a smaller volume, increasing nonlinear effects.

5. Your system has to transmit wavelength-division multiplexed signals at the 1530 to 1620 nm band of erbium-doped fiber amplifiers. What type of fiber is best?

Answer: Non-zero dispersion-shifted fiber. Depending on the details, you can use reduced-dispersion-slope or large-effective area fiber with the same properties.

6. Your cheapskate purchasing department just got a great deal on zero dispersion-shifted fiber. Why can't you use it in the erbium-amplifier system in Question 5?

Answer: Zero dispersion-shifted fiber suffers from strong nonlinear effects at 1550 nm which causes four-wave-mixing crosstalk between channels.

<u>Worked Quiz Problems</u>

Problem 1:

First calculate the confinement angle

$$\theta_{confinement} = \arccos\left(\frac{n_{clad}}{n_{core}}\right) = \arccos\left(\frac{1.495}{1.5}\right) = 4.678°$$

Then calculate the half acceptance angle from that

$$\theta_{accept} = \arcsin\left(n \times \sin\theta_{confinement}\right) = \arcsin(1.5 \times 0.0158) = 7.029°$$

Students who forget that they need to account for the difference between confinement and acceptance angle will get the wrong answer.

PROBLEM 6:

This is the case of a bare glass fiber in air. Use the formula that gives maximum core diameter for a single-mode step-index fiber, plugging in the proper numbers for refractive index and wavelength:

$$D < \left(\frac{2.4\lambda}{\pi\sqrt{n_0^2 - n_1^2}}\right) = \left(\frac{2.4 \times 1.3\mu m}{3.1416\sqrt{\left(1.5^2 - 1.0003^2\right)}}\right) = 0.89\mu m$$

This shows that an unclad fiber would have to be impractically small to transmit a single mode.

PROBLEM 7:

Use the same formula for single-mode core diameter as in Problem 6, but with different numbers:

$$D < \left(\frac{2.4\lambda}{\pi\sqrt{n_0^2 - n_1^2}}\right) = \left(\frac{2.4 \times 1.3\mu m}{3.1416\sqrt{\left(1.5^2 - 1.495^2\right)}}\right) = 8.1\mu m$$

Single-mode core diameter increases as the refractive index of the cladding approaches that of the core, because you divide by the square root of the difference of the squares. This is an important element in fiber design.

<u>Answers to Quiz for Chapter 4</u>

1. b

2. a

3. e

4. e

5. c

6. b

7. d

8. c

9. b

10. a

11. d

12. c

Supplementary Material

Cross-sectional diagrams of different fiber types (e.g., Figure 4.1 and Figure 4.11).

Plots of dispersion versus distance for various fiber types.

Computer routines to calculate cumulative dispersion versus distance, which yield graphs you can show at lectures. Spreadsheets are convenient, especially for plotting the results in graphical form.

For more details on fiber types and design, see Gerd Keiser, *Optical Fiber Communications,* 3rd ed (McGraw-Hill, 2000)

Chapter 5:

Properties of Optical Fibers

This chapter covers the most important fiber properties: attenuation, light collection, dispersion, nonlinearities, and strength. Other aspects of some of these topics are covered in other chapters, including dispersion in Chapter 4 and light collection in the description of splices and connectors in Chapter 13. This chapter also explains calculations on a decibel scale, the most convenient way of measuring attenuation.

<u>Overview</u>

Attenuation measures the amount of light lost between input and output. It includes light never collected by the fiber, light absorbed by the fiber material, and light scattered from the fiber core so it never reaches the output. I describe these contributions and put them together in a single equation. Then I look separately at attenuation and light coupling effects.

Figure 5.2 illustrates the contributions of the main elements of attenuation: Rayleigh scattering by atoms in the glass at short wavelengths, infrared absorption of silica at long wavelengths, and impurity absorption peaks. The plot does not represent any particular fiber, and the scale tends to exaggerate the impact of the peaks, but it's useful in pointing out the components.

The discussion of attenuation also elaborates on decibel calculations, a topic introduced in Chapter 2 which is critically important for students to understand. Appendix B lists decibel equivalents on other scales. This section points out the use of decibel units and calculations in system design.

The two key elements of light collection covered here are core size and numerical aperture. Although NA is not particularly useful in working with the single-mode fibers that are standard for telecommunications, it is important for multimode fibers used for short-distance communications, illumination, and imaging. This section also mentions other source of loss, particularly leaky modes and bending losses.

Chapter 4 described the importance of fiber dispersion. In this chapter, I step readers carefully through the different types of dispersion and how they add to give total dispersion. There is a bit of algebra here, but it's simple and critical. You need to stress that dispersion adds by a sum-of-squares law, not by simple linear addition, so that the larger dispersion dominates. You also should stress that the magnitude of chromatic dispersion depends on source bandwidth as well as fiber properties, unlike pure modal dispersion. Figure 5.9 gives graphical display of how the variation of refractive index with wavelength gives rise to material dispersion. It may be worth walking students through this chart. I also introduce polarization mode dispersion. The section concludes with formulas showing the dependence of data rate and bandwidth on total dispersion, a topic revisited in Chapter 21 on system design.

Fiber nonlinearities also impose limits on system performance at high speeds, so I describe the major types and their impact. Their strength increases greatly with power carried in the fiber, which particularly impacts high-speed systems because they must carry high average power, and WDM systems because they carry optical power at many wavelengths.

The description of mechanical properties is comparatively brief because students generally have been exposed to such mechanical concepts, and quickly understand that fibers will break if they're overstressed.

Questions to Think About for Chapter 5

1. The amount of Rayleigh scattering by atoms is proportional to λ^{-4}. How is this related to why the sky looks blue?

Answer: The color of the sky comes from scattered sunlight. Blue wavelengths are at the short end of the visible spectrum, so they are most likely to be scattered by atoms and molecules in the atmosphere. The amount of scattering also depends on the viewing angle and the path between your eye and the atoms that scatter the light. Violet light is shorter in wavelength, but our eyes are much less sensitive to it. (For a fuller description, see Robert Greenler, Rainbows, Haloes, and Glories *(Cambridge University Press, 1980).)*

2. Can you write a formula that converts loss in decibels into the fraction of power remaining?

Answer: Recall the formula for loss in decibels is:

$$dB\,loss = -10\log_{10}\left(\frac{P_{out}}{P_{in}}\right)$$

You want to solve for the ratio of power output to power input. Doing so gives (in steps)

$$\frac{-dB\,loss}{10} = \log_{10}\left(\frac{P_{out}}{P_{in}}\right)$$

then

$$\left(\frac{P_{out}}{P_{in}}\right) = 10^{\left(\frac{-dB\,loss}{10}\right)}$$

3. You have a fiber that transmits a single mode at 850 nm, and a light source with bandwidth of 1 nm. Its chromatic dispersion is about -80 ps/nm-km. What is the maximum data rate that fiber could transmit 100 km, neglecting attenuation, based on the guideline for chromatic dispersion (bit rate = $1/(4 \times$ dispersion))?

Answer: The total chromatic dispersion is 100 km × -80 ps/nm-km, or 8 ns for a 1nm source. That corresponds to maximum data rate of 31 Mbit/s. This was a major reason single-mode fiber was never very attractive at 850 nm.

4. Estimate the attenuation at 850 nm from Figure 5.2. Assume you need an optical amplifier to boost signal strength after every 30 dB of fiber loss. How far can the fiber transmit signals before it requires an optical amplifier? Which of the limitations in Questions 3 and 4 do you think was the main reason 850-nm systems were never viable for long-distance transmission?

Answer: Loss is about 2 dB/km, so an optical amplifier is needed every 15 km. The data rate limitation is more serious, because optical amplifiers cannot eliminate accumulated dispersion effects. The only long-distance systems ever used at 850 nm required repeaters about every 10 km that completely regenerated the signals. Optical amplifiers were not then available.

5. Why is it more difficult to compensate for dispersion in a DWDM system than in one transmitting only a single optical channel?

Answer: Dispersion slope. With two fibers having chromatic dispersion of different sign, you can always come up with some combination that will make the net dispersion any low number you want. However, because fiber dispersion varies in different ways with wavelength, any combination of fibers that works for one wavelength generally won't work for others.

6. Write a formula for how long a length of dispersion-compensating fiber ($L_{disp\text{-}comp}$) you need to add to an existing system with $L_{existing}$ km of fiber to reduce dispersion to a desired value D_{net}.

Answer:

$$D_{net} \times \left(L_{existing} + L_{new} \right) = D_{existing} L_{existing} + D_{disp\text{-}comp} L_{disp\text{-}comp}$$

7. Four-wave mixing normally occurs only at high powers in glass, yet it can cause significant crosstalk in single-mode fibers. If you have a fiber with effective area of 50 μm^2, what is the power per square centimeter in the fiber if it carries 100 optical channels at 3 mW each?

Answer: 600,000 watts per square centimeter. The total power is only 0.3 W, but the effective area is only 0.0000005 square centimeter.

Answers to Quiz for Chapter 5

1. e

2. d

3. d

4. c

5. a

6. b

7. a

8. c

9. c

10. b

11. e

12. d

13. c

14. b

<u>Solved Problem</u>

PROBLEM 12.

Fiber strength is expressed in pounds per square inch of glass cross-sections, so first calculate the cross-sectional area of the glass.

$$A = \pi r^2 = \pi(0.0625\ mm)^2 = \pi(0.00246\ in)^2 = 1.902 \times 10^{-5}\ in^2$$

Multiplying this small area by the large strength per unit area shows the fiber can support 11.4 pounds, the weight of a rock or small concrete block, not an elephant, but respectable for something that thin.

<u>Supplemental Material</u>

Tables of decibel equivalents in Appendix B, or easily produced with a spreadsheet.

Color versions of graphs dividing fiber attenuation between absorption and scattering. (You can blow up Figure 5.2 and add color.)

Computer routines to calculate attenuation and dispersion, for classroom display. Pick representative values and show students the differences.

Plots showing transmission speeds over various distances possible with different levels of dispersion.

Chapter 6:

Fiber Materials and Manufacture

This chapter covers the materials used to make optical fibers, as well as manufacturing techniques for glass-based fibers. Plastic and other exotic fibers require specialized production techniques not covered here. You should stress how fiber characteristics depend on the choice of materials, including their purity.

<u>Overview</u>

Most optical fibers are made of glass, so I concentrate on glass fibers and the techniques used to make them. Note the importance of explaining what people mean when they say "glass." In the fiber-optic community, "glass" generally means a clear solid rich in silicon oxide or silica, but more generally it means a disordered solid that may not necessarily be clear. In practice, the characteristics of glass-core fibers can vary widely, and some types do not have the extreme purity needed for telecommunications. I inserted the box showing that glass does not flow at room temperature because this is widely misunderstood and interesting.

Fiber fabrication is a process vital to the industry, but most students need not worry much about it in practice. Explaining fiber fabrication both answers questions from the curious and helps explain some limitations on the choice of fiber materials. The rod-in-tube process is simple to explain, but is used only for imaging and illumination fibers. Vapor deposition processes are needed for low-loss communication fiber. This section covers fused silica fibers, and does not cover all variations of the usual processes.

The dichotomy between glass and plastic remains a major one in fiber optics, and it is essential that students understand it. Throughout most of this book, I concentrate on core and cladding composition, but you should remind them that most glass-core fibers also have an outer plastic coating or buffer layer for mechanical protection. Some of these coatings are color-coded. Although plastic cladding of glass fibers sounds very simple, achieving the required surface quality often is difficult, and glass-clad glass-core fibers dominate except in some large-core illuminating fibers.

All-plastic fibers have been around for many years, but have found only limited applications because of their high attenuation and limited tolerance of high temperature. Most are step-index multimode, but recently graded-index multimode fibers have become available.

Liquid-core light guides, mid-infrared fibers, and hollow optical waveguides are described briefly for completeness. They are not used for telecommunications. Most students will not encounter them, and don't need to know much about them—but they should know what they are.

"Holey" or microsctructured fibers are a hot research topic, covered briefly here as a type of general transmission fiber, and in Chapter 7 as a type of specialty fiber. Their internal structures come from internal microstructures. Specialists call the general class of fibers "photonic crystal" fibers because some types guide light in ways that do not depend entirely on a photonic bandgap. You need not worry much about this, but these fine distinctions do contribute to inconsistencies in the terminology used in research papers.

The concept of the planar optical waveguide is treated here because it's another fundamental type of optical waveguide which becomes important in later chapters. You should point out that planar waveguides rely on the same principles of light guiding as optical fibers, but have a different geometry.

Questions to Think About For Chapter 6

1. Window glass looks very clear when you look straight through a pane, but when you look into the edge it looks green. What causes this color?

Answer: Impurities in the glass. Window panes looks very clear because we only look though about 2 mm. If you look from edge to edge of a small pane, the glass is about 20 cm thick. If the glass had 0.05 dB/mm absorption in the green, you would see only 0.1 dB of attenuation through the pane, but 10 dB looking edgewise. Most losses in thin windows actually are caused by surface reflection.

2. Why is it easy to make fibers from glass but impossible to make them from ice, even in Antarctica?

Answer: Glass molecules hold tenuously to each other, forming a viscous liquid just above the melting point. As you pull molten glass into a fiber, it quickly cools to form a solid. Water molecules hold each other much more weakly, so molten ice is the thin watery liquid we call water.

3. What are the main trade-offs in picking dopants for the core and cladding of fused silica fibers?

Answer: Dopants for the core must have very low absorption, or they increase attenuation in the core. Pure silica would have very low absorption for the core, but the cladding must have a lower refractive index, The only dopants that can reduce the index of fused silica are fluorine and boron. Germanium is the only dopant that raises core index without inducing excess absorption or other undesirable side effects. Only a small amount has to be added because the core index typically is only 0.3 to 1% higher than the cladding index.

4. The large-core step-index fibers listed in Table 6.1 have cores that are nominally pure silica and generally carry higher laser powers than telecommunication fibers. Yet their attenuation is higher than for single-mode telecommunication fibers at the same wavelengths. Why should this happen?

Answer: The silica is not purified to the same extent as in telecommunication fibers which have to carry signals tens or hundreds of kilometers. Large-core step-index fibers are used for beam delivery over distances of meters or less, so their cumulative attenuation is very low. The small volumes of silica used in the cores of single-mode fibers can be purified to much higher levels. A 9-μm single-mode core has a cross-sectional area of 64 square micrometers. A 200-μm core fiber has cross-sectional area of 31,000 μm^2, nearly 500 times larger.

5. How would you compare the advantages of plastic and glass fibers? What are the best features of each? Where might plastic have an advantage?

Answer: Plastic fibers are easier to handle and more flexible, making installation cheaper; glass fibers have much lower attenuation and can withstand higher temperatures and generally harsher operating conditions. Plastic fibers are attractive where they need only span short distances in relatively protected environments and must be handled inexpensively or by technicians not familiar with connecting glass fibers. Think of a do-it-yourself auto mechanic trying to repair a broken fiber.

6. Roughly how many times higher is the minimum loss of plastic fibers than glass fibers, measured in dB/km?

Answer: Over 250 times, the clearest glass fibers have loss of 0.2 dB/km near 1550 nm. The best plastic is around 50 dB/km.

7. If you aimed a 1.55-μm laser beam straight up through clear air, about 90% of the light would escape into space, with 10% scattered or absorbed. The atmosphere becomes more tenuous at higher altitudes, so assume that sending a beam into space is equivalent to transmitting it through 10 km of air. What is the equivalent attenuation in dB/km? How does this compared with the best optical fibers at that wavelength? Does that imply a limit on the transparency of hollow photonic bandgap fibers?

Answer: 90% transmission corresponds to a loss of 0.46 dB, or attenuation of roughly 0.046 dB/km. The clearest commercial fibers have loss about 0.2 dB/km at 1.55 μm, with laboratory fibers going to 0.15 dB/km. If hollow fibers were filled with clear air, neglecting any waveguide losses the maximum transmission distance would be only 4.3 times that for 0.2 dB/km glass fibers. However, the loss of 'clear' air includes some scattering and absorption by dust and aerosols which would be unlikely to remain in the air in a hollow fiber. Sealed hollow fibers could be kept at low pressure or filled with a relatively clear gas such as nitrogen. The major advantage of air is that is has extremely low chromatic dispersion at wavelengths longer than 1 μm.

Answers to Quiz for Chapter 6

1. a

2. d

3. e

4. c

5. e

6. b

7. a

8. b

9. a

10. c

11. c

12. a

13. c

14. a

Supplemental Materials

Samples of non-transparent and colored glass to show the diversity of materials called "glass."

Glass rod or tube to heat in flame and stretch out into a (not particularly clear) fiber.

Samples of other transparent or semi-transparent materials, such as plastics.

Microstructured fibers are hot research topics, and much current information is available on the web. The best approach is to search for "photonic crystal fiber" and "photonic bandgap fiber". Much of the work is performed in Europe, so be sure to try searching with "fibre" substituted for "fiber."

Chapter 7:

Special-Purpose Fibers

This chapter covers fibers designed for purposes other than simply transmitting light. These include dispersion, compensating fibers, polarization-maintaining fibers, fiber Bragg gratings and doped fibers used in optical amplifiers. This introduction to fiber gratings is important to understanding the later description of fiber gratings for WDM in Chapter 15. The description of doped fibers introduces the concept of optical excitation covered in more detail in the section on lasers in chapter 9. Fiber amplifiers *per se* are described in Chapter 12.

Overview

Specialty fibers are designed to have properties which meet specific requirements which go beyond the normal range of transmission applications. The opening section briefly describes several types of practical importance:

- Dispersion-compensating fibers

- Polarization-maintaining fibers

- Bend-insensitive and coupling fibers

- Reduced-cladding fibers

Doped fibers have become important for use in optical amplifiers and fiber lasers. Their operation depends on a light-emitting impurity element added to the fiber core, which is excited with an external light source. The dopant used in most doped fiber amplifiers is erbium, a rare-earth metal which emits in a band centered near 1.55 μm. Other dopants can work in other windows, but none are as effective. This chapter briefly explains the concept, but concentrates on the designs used for the specialty fiber. Chapter 9 covers the laser principle in more detail; Chapter 12 covers optical amplifiers.

Fiber gratings are versatile optical components made by altering the refractive index of selected regularly spaced zones within the cores of fibers. Light scattering from these high-index zones reflects the wavelength selected by the grating period, while transmitting other wavelengths. Their major impact so far has been in optics for wavelength-division multiplexing. This section also describes how gratings can be placed in series to selectively reflect multiple wavelengths, or can be "chirped," which spacing changing along the fiber length, for dispersion compensation.

This chapter closes with sections covering photonic crystal or "holey" fibers and special fibers developed for optical displays and imaging. It does not cover sensing fibers, which are the subject of Chapter 29.

Questions to Think About for Chapter 7

1. Reduced-cladding fibers have outer diameters of 80 μm, but developers have proposed 60 μm claddings. That should not affect optical performance because the claddings would be about 25 μm thick. What problems might be expected from the smaller diameter?

Answer: Handling would be more difficult, because the finer fibers are more flexible, and harder to grasp, to feel, and even to see.

---

2. A FIBER GRATING reflects light waves that are in phase with the grating, with a wavelength $\lambda = 2nD$ where D is the spacing between the gratings. This wavelength is the longest that would make exactly one round trip between a pair of grating lines. A wavelength exactly half that value also would be resonant with that grating spacing, because exactly two waves would make one round trip. Why isn't that light also reflected?

Answer: That light is not in the signal. WDM signals typically are in the range of 1525 to 1625 nm. One-half of 1550 nm (for example) is 775 nm. These different resonances are analogous to the multiple orders of a diffraction grating, which we don't see overlap because our visual range is limited.

3. Why is refractive index always measured in air rather than in glass?

Answer: Glasses have different values of refractive index, so you would have to specify which glass. Historically, wavelengths have been calibrated by experiments performed in air. Precise wavelengths are calibrated as if light was in a vacuum, but the refractive index is almost the same in air, which at standard temperature and pressure has a refractive index of 1.000293. The refractive index varies slightly with temperature and pressure, but vacuum wavelengths are the same everywhere.

4. What's the advantage of doping erbium only in the fiber core?

Answer: The input signal is directed into the core of the amplifying fiber, so all the signal starts there. Emission stimulated by light guided in the fiber is emitted in the same direction as the input signal, keeping the amplified signal in the core. Moreover, erbium ions in the cladding could absorb pump light but would not contribute significantly to the amplified signal, reducing total efficiency.

5. The gain of an erbium-doped fiber varies as a function of wavelength. It is highest from about 1530 to 1565 nm, and lower at longer wavelengths. Recalling how gain is defined, what is one way to make an erbium-doped fiber amplifier with a high amplification factor at longer wavelengths?

Answer: Remember that gain is the amplification per unit length. The most obvious way to increase total amplification is to use a longer fiber. In practice, pump power and erbium doping density also has to be adjusted.

6. Think of a way to demonstrate the effect seen in light scattering from the sides of a fiber designed for decorative lighting. Start with a laser pointer.

Answer: The simplest approach is to aim the laser pointer through a dusty zone (chalk dust should work well). Light scattered off the dust makes the beam visible. If the air was perfectly clear, no light would be scattered, and you wouldn't see the beam in the air. To show a beam in water, add a dash of milk to otherwise clear water; milk solids will show similar scattering.

7. An erbium-doped fiber amplifier is operating at a power high enough that its output has saturated at 10 mW/channel for 20 optical channels with 100-GHz spacing. You add 20 more channels at intermediate wavelengths. Assuming the amplifier is so saturated it can't generate any more total output, what is the new output per channel?

Answer: One-half the original 10 mW/channel, or 5 mW/channel. Total output is the same, but there are twice as many channels, so there's only half as much output power per channel.

Answers to Quiz for Chapter 7

1. b

2. b

3. d

4. c

5. d

6. c

7. e

8. a

9. d

10. a

Supplemental Materials

Illustrations from other chapters (e.g., Chapter 12) showing fiber amplifiers in place in a transmission system.

Illustrations of the principles of a solid-state crystalline or glass laser, which are similar to those of fiber amplifiers.

Plots of light transmission and reflection by fiber gratings (see Chapter 15).

Chapter 8:

Cabling

Fibers must be cabled for use in practical communication systems. This chapter outlines the theory and practice of cabling, assuming the students will be using but not designing cables. The chapter is inherently more descriptive than quantitative. You can think of it as a field guide to cabling, to help students identify cable types, their structures, and their uses. It is not intended to teach installation procedures.

<u>Overview</u>

Cabling techniques developed for copper wires were transferred to fiber optics when serious development started on fiber communications. Unless a cable is clearly marked or close enough for you to handle or see a cut end, it may be hard to tell if it contains wires or optical fibers.

I open by explaining the reasons for cabling—making fibers easier to handle and protecting them from damaging forces and the environment. It's easy to make this point by having students handle bare fibers and sample cables.

The next section covers the types of fiber-optic cable. This is organized largely by the environmental and operational requirements, which are the dominant influences on cable design. You should note that fire and electrical safety standards determine what types of cables may be installed indoors.

Note that the choice of cabling depends on the environment in ways that may not be obvious. For example, the heaviest armored cables are not those laid at the bottom of the ocean, but those buried just under the ocean floor in shallow water, where they must withstand the assaults of fishing trawlers and ship anchors. You may have heard reports of sharks attacking some undersea cable; that incident occurred in the mid-1980s, and apparently was due to sharks sensing an electric field from the current flowing through the cable to power repeaters.

Gophers gnawing on cables may sound funny, but it's a serious business for companies in areas where gophers live. If you want a chuckle from your class, you can mention that Bellcore (now Telcordia Technologies) once had a standard for gopher-testing which involved stretching a piece of cable through the cage of a hungry gopher. Cables containing obsolete fibers are sometimes called "gopher bait."

The following section describes the functional elements that go into cables, and makes important distinctions among cable structures. The black and white illustrations in the book cannot illustrate the color coding of fibers or cable substructures, which is important in cable management.

The final brief sections cover cable installation and cable changes and failure.

<u>Questions to Think About for Chapter 8</u>

1. The design of fiber-optic cables evolved from the design of electrical cables for similar applications. What is one common problem with aging electrical cables that does not affect fiber-optic cables?

Answer: Short circuits between conductors because of insulation breakdown or moisture in the cable. Another is electronic crosstalk between parallel conductors.

2. What is one problem with fiber-optic cables not present with electronic cables?

Answer: Microbending loss in the fiber. This problem can be greatly reduced by choice of cable structure.

3. Why make one cable with 1000 fibers rather than 10 with 100 fibers?

Answer: Installation and right-of-way costs are much lower for one cable than for 10.

4. A cable weighs 300 kilograms per kilometer, with poles placed 50 meters apart on the road. A flock of 20 0.5-kilogram crows perch on the cable to watch the scenery. How much do they increase the weight of the cable span, in percent?

Answer: At 0.3 kg/m, a 50-meter length of cable weighs 15 kilograms. The crows add 10 kilograms, increasing the weight 67%.

5. An ice storm coats the cable with a one-centimeter layer of ice. Assume for simplicity that the cable is 2 cm in diameter, and that ice weighs 1 gram/cubic centimeter. How much weight does the ice add to a 50-meter span of the cable?

Answer: The cross-sectional area of the ice is an annulus. Use a 1-cm radius for the cable and adding a 1-cm layer for the ice. Calculate the difference between the area of the cable and that of the ice-coated cable.

$$Area = \pi \left[(2\ cm)^2 - (1\ cm)^2 \right] = 9.42\ cm^2$$

The volume of ice is 9.42 cm^2 times the 50-meter length, or 47,000 cm^3, which comes to 47 kg, three times the 15-kg weight of the cable.

6. What is the principal difference between the structures of indoor and outdoor cables?

Answer: Physically, loose tube cables are used more outdoors, while tightly buffered cables are used mostly indoors. Tightly buffered cables generally meet fire codes better. They also are designed for stable-temperature environments. Loose-tube cables typically can better withstand temperature changes.

<u>Answers to Quiz for Chapter 8</u>

1. c

2. c

3. a

4. b

5. c.

6. a

7. c

8. b

9. a

10. d

<u>Supplemental Materials</u>

Collect a few cable catalogs and make slides of the color pictures showing different types of cable. You can find images on the web sites of many cable manufacturers.

Sample lengths of different cables vividly show students the differences among them.

Get samples of various types of cables and open them up to see their internal structure.

Show videos demonstrating different type of cable installation.

Chapter 9:

Light Sources

$\mathbf{L}$ ight sources are vital components in all fiber-optic systems. This chapter describes the light sources used in virtually all fiber-optic communication systems: LEDs, semiconductor lasers, and fiber lasers (optical amplifiers are covered in Chapter 12). Students should learn how they operate, their performance in fiber-optic systems, and their major differences. The transmitters built around light sources are covered in Chapter 10. You should cover the two chapters in sequence; I split them to keep each chapter a reasonable length.

<u>Overview</u>

The chapter opens by discussing the major considerations in selecting fiber-optic light sources: wavelength, output power, modulation capability, and cost. Designers balance these considerations in picking light sources for particular applications. There is no one "right" light source for all fiber-optic systems.

LEDs are a logical starting point because they are the simplest of the semiconductor light sources. I describe their operation as forward-biased semiconductor diodes, which generate light by recombination of electrons and holes at the junction of p- and n-type materials. I cover only those details of semiconductor physics essential to understanding semiconductor lasers based on compound semiconductors such as GaAs, so you may get questions from students curious about silicon and electronic components.

Next I introduce the principles of lasers in general, including population inversion, excitation, amplification and oscillation. This is no substitute for a good laser course, but it will prepare lasers to learn what they need to know about semiconductor lasers for fiber-optic applications.

Semiconductor diode lasers (often called simply laser diodes or diode lasers) are covered next, starting with simple stripe-geometry Fabry-Perot lasers. My general description covers light confinement and beam generation in edge-emitting lasers and compares them to LEDs. A separate section covers vertical cavity surface-emitting lasers (VCSELs) because their operation is so different than edge emitters. Another section covers the materials used in diode lasers, their output wavelengths, and their spectral properties, all important properties for fiber-optic systems. This section covers single-frequency and tunable lasers, as well as the effects of direct modulation of diode lasers.

This chapter closes by explaining operation of fiber and solid-state lasers, which have some applications in fiber-optic systems. This section assumes the student has read about doped fibers in Chapter 7.

<u>Questions to Think About for Chapter 9</u>

1. Below threshold a diode laser emits some light by spontaneous emission. Why does it behave like an LED?

Answer: Because below laser threshold it is an LED. Current levels have not built up enough recombination to produce a population inversion, so the laser is operating in LED mode. Different devices are used as LEDs and lasers because their designs must be optimized in different ways for spontaneous emission and to withstand the high current densities of lasers.

2. A diode laser has a threshold current of 12 milliamperes. That current passes through a stripe 5 µm wide and 500 µm long. What is the current density in amperes per square centimeter?

Answer: The stripe area is 2500 $µm^2$, or 0.000025 cm^2. Dividing that area into the threshold current gives a threshold current density of 480 A/cm^2. Current density is high because it is concentrated in a small area, and because it's measured per square centimeter, which is much larger than the area of the device.

3. The diode laser in problem 2 emits 5 mW of light. The junction layer is 0.5 µm thick. If the light is evenly distributed across the end of the active layer, what is the optical power density in W/cm^2?

Answer: 200,000 W/cm^2. The emitting area is 5 µm wide by 0.5 µm thick, so output is 2 $mW/µm^2$. As with current density, the number is high because the area is very small.

4. InGaAsP has a refractive index of about 3.5. A Fabry-Perot laser emitting multiple longitudinal modes at a nominal wavelength of 1550 nm has a cavity 500 µm long. What is the wavelength difference between two adjacent longitudinal modes.

Answer: Calculate two modes using the formula 2Ln=Nλ for a cavity L long.

Start with the 1.55 µm wavelength to estimate a value of N.

$$N = \frac{2Ln}{\lambda} = \frac{2 \times 500 µm \times 3.5}{1.55 µm} = 2258.06$$

This isn't an integer, so you need to calculate the wavelength for the closest integers, N=2258 and 2259.

$$\lambda = \frac{2Ln}{N} = \frac{2 \times 500 µm \times 3.5}{2258} = 1.55004 µm$$

and

$$\lambda = \frac{2Ln}{N+1} = \frac{2 \times 500 µm \times 3.5}{2259} = 1.54936 µm$$

which indicates the difference is about 0.00068 µm, or 0.68 nm. Rounding errors occur if you try to calculate the offset from 1.55 µm.

5. Suppose the laser in problem 4 was a VCSEL with cavity length L of only 10 µm, and for the time being forget about the difficultly of making InGaAsP VCSELs. Calculate the separation between two modes using the same technique.

Answer: Calculate two modes using the formula $2Ln=N\lambda$

Start with the 1.55 µm wavelength to get a value of N.

$$N = \frac{2Ln}{\lambda} = \frac{2 \times 10\mu m \times 3.5}{1.55\mu m} = 45.1613$$

Then go back and calculate wavelengths for N=45 and 46

$$\lambda = \frac{2Ln}{N} = \frac{2 \times 100\mu m \times 3.5}{45} = 1.55556\mu m$$

and

$$\lambda = \frac{2Ln}{N} = \frac{2 \times 10\mu m \times 3.5}{46} = 1.52174\mu m$$

which shows a much larger difference of 33.82 nm.

This answer shows how a short cavity can limit a laser to oscillating in a single longitudinal mode by pushing other adjacent modes outside the laser's gain band.

<u>ANSWERS TO QUIZ FOR Chapter 9</u>

1. a

2. b

3. e

4. a

5. a

6. b

7. d

8. b

9. c

10. c

<u>Worked Quiz Problem</u>

Problem 12.

It takes a couple of steps to calculate the wavelengths of two adjacent longitudinal modes. The first is to estimate mode numbers (which are not exact values). The second is to calculate the wavelengths for modes N and N+1.

The number of modes in a 500-μm cavity of material with refractive index 3.2 at a wavelength of 1.3 μm is approximately

$$N = \frac{2Ln}{\lambda} = \frac{2 \times 500\mu m \times 3.2}{1.3\mu m} = 2461$$

Then flip the equation around and calculate wavelengths for modes 2461 and 2462.

$$\lambda = \frac{2Ln}{N} = \frac{2 \times 500\mu m \times 3.2}{2461} = 1.3002844\mu m$$

$$\lambda = \frac{2Ln}{N} = \frac{2 \times 500\mu m \times 3.2}{2462} = 1.299756\mu m$$

Subtracting gives a wavelength difference of 0.000528 μm or 0.53 nm.

Supplementary Material

A RED LASER POINTER is an example of a diode laser with a wavelength used for plastic fibers. You should note that the lasers used with glass optical-fiber communication systems emit invisible infrared light. CD players use 780-nm lasers, which also are invisible to the eye, but are not as long as the standard fiber windows at 850, 1300, and 1550 nm.

A periodic chart of the elements shows the relationship of the III-V elements used in semiconductor lasers and LEDs.

A spectral chart will let you highlight important wavelengths.

BOOKS ON LASERS GIVE more details on laser operation. See, for example, Jeff Hecht, *Understanding Lasers* (Wiley/IEEE Press, 1994), or Breck Hitz, J. J. Ewing and Jeff Hecht, *Introduction to Laser Technology* (Wiley/IEEE Press, 2001)

Chapter 10:

Transmitters

Transmitters take an input electrical signal and process it so it can drive a light source that emits an optical signal transmitted through a fiber-optic system. This chapter describes how transmitters work, assuming students have learned about light sources in Chapter 9. It covers operational considerations, multiplexing, modulation, and transmitter design, and includes some elements of overall communication system design.

Overview

This chapter opens with a brief description of transmitter concept. Fiber-optic transmitters are inevitably hybrids, part optics, part electronics, and part communications equipment. The overall performance of a transmitter depends on properties of the light source, the drive electronics, and the overall communication system.

The next section covers important operational considerations for fiber-optic transmitters, including brief descriptions of packaging, performance measurements, digital and analog transmission, and ways of measuring bandwidth. Illustrations show a couple of standard transmitter modules.

The following section briefly explains all three types of multiplexing that may be encountered in fiber-optic transmitters: time-division multiplexing that interleaves bit streams, frequency-division multiplexing that combines analog signals, and wavelength-division multiplexing that sends multiple wavelengths through a single fiber. Diagrams show their operation.

The section on modulation starts by explaining the amplitude or intensity modulation that is standard in fiber-optic systems. Figure 10.6 shows the important concept of direct modulation of a laser source by combining a bias current set slightly above laser threshold with a modulation current that is added to raise "on" pulses to higher power. A brief subsection explains the idea of coherent transmission, drawing an analogy with FM radio.

The final two sections give an overview of what goes into transmitter design, with few details. I treat transmitter design primarily at the block-diagram or functional level. At this level, students should learn *what* transmitters do, but they need now worry exactly *how*. Thus I give only a couple of very simple circuit diagrams at the end of the chapter, but I describe important circuit functions and show what packaged transmitters look like. You may want to add examples, particularly transmitters you use in the laboratory.

<u>Questions to Think About for Chapter 10</u>

1. The difference between the "off" and "on" states of a laser transmitter is 20 dB. The "on" output is 1 mW. What is the "off" output at the output port of the transmitter? What is the "off" output after 30 dB of loss?

Answer: Off output is 10 mW (-20 dBm) at the output port and 10 μm (-50 dBm) after 30 dB loss.

2. The diode laser used in a transmitter has a threshold current of 10 mA and normally operates at 15 mA in the "on" state. Suppose it is directly modulated with a signal without a bias current, which turns the laser on from zero current. The signal pulse has 30-ps rise time, 30-ps "on" time, and 30-ps fall time. How long is the output optical pulse, starting from the time the current crosses laser threshold.

Answer: 50 ps. The optical pulse turns on 20 ps into the 30-ps rise time, when current reaches 10 mA. It remains on until 10 ps after the electrical pulse starts dropping, when current drops below threshold. This shortens a 90-ps pulse to 50 ps.

3. Suppose you could find electronics fast enough to generate time-division multiplexed signals at 640 Gbit/s on a single optical channel. How many light waves would one bit correspond to at a wavelength of 1550 nm?

Answer: One bit at 640 Gbit/s would last 1.56 picosecond. The frequency of a 1550-nm signal is 193.548 THz. Multiplying by the pulse duration gives 302 waves per bit.

4. Time-division multiplexing requires reducing the duration of low-speed pulses and interleaving them into a combined signal. It now is done electronically. Suppose you had a way to reduce the duration of optical pulses. Can you think of a way to interleave these shortened optical pulses for time-division multiplexing? Consider a system that multiples data rate by a factor of four. (Hint—think about ways to delay pulses.)

Answer: Insert lengths of optical fiber that provide the required delay. One input would enter in normal phase, a second would be delayed one bit interval, a third two bit intervals, and the fourth three bit intervals. This is done in laboratory experiments where data rates are faster than commercial electronics.

<u>Quiz for Chapter 10:</u>

1. d

2. a.

3. c.

4. a.

5. c

6. c.

7. a.

8. d.

9. d.

10. b.

11. c.

12. c.

13. b.

14. d.

<u>Supplemental Material</u>

Bring in a real transmitter and open the box to show which components serve particular functions. Do this for two or more different models to show the differences among fiber transmitters. Photos or good diagrams suffice for other types.

Obtain samples of standard laser modules and explain how they work. The groups that established the standard formats may have cut-away diagrams available on their web sites.

Point out the difference between the small light source module with drive electronics packaged in a DIP-type package, and a rack-mounted complete transmitter.

Compare transmitters with transceivers, which include receivers as well as transmitters.

Show a chart of the broadcast spectrum for radio, or frequency allocations for television channels to illustrate the idea of frequency-division multiplexing.

Appendix D lists the ITU wavelength channels. You can compare these to standard radio frequencies in the AM or FM bands, or standard television channel slots.

Chapter 11:

Receivers

Receivers convert optical signals into electronic form for use by other devices or communication systems. This chapter describes both receivers and the detectors within them that actually sense optical signals. As with transmitters, I emphasize the optical and communication-system aspects rather than the electronics.

Overview

This chapter opens with a definition of receivers. A block diagram illustrates the basic elements that may be included in a receiver: an optical preamplifier, a wavelength-division demultiplexing stage, a detector, an electronic amplification stage, and circuits that clean up the final signal in digital receivers. Most receivers are packaged without the optical preamplifier or demultiplexing stage, but those stages conceptually are part of a receiver. Where demultiplexing is required, separate receiver modules are needed for each wavelength channel.

The detectors used in fiber-optic receivers are semiconductor diodes that are reverse-biased so they do not carry a current in the dark. When light reaches the diode junction, it produces *p-n* pairs, generating a photocurrent. The wavelengths to which detectors respond depend on the choice of material, with response curves shown in Figure 11.5. The speed and sensitivity of a photodetector also depend on the design of the photodiode. I describe several types, and summarize their properties in Table 11.2. The most widely used are *pin* photodiodes, with avalanche photodiodes also important.

The section on performance considerations concentrates on detectors, but the concepts also apply to receivers. I describe the difference between responsivity and quantum efficiency, and introduce dark current and noise-equivalent power. The description of speed and bandwidth introduces the concepts of rise and fall time. I also cover signal coding, quality, and power.

The section on electronic functions breaks them into several categories. Earlier I broke them into two broad categories: analog functions such as amplification, and digital functions such as retiming and regeneration. In that sense, all digital receivers contain internal analog stages that perform amplification. Equalization and filtering are other "analog" functions. I separate discrimination and timing as fundamentally digital processes.

The final section gives a few simple circuit diagrams to illustrate the electronic design of receivers.

Questions to Think About for Chapter 11

1. THE INPUT SIGNAL at a receiver is -30 dBm (1 µW), which is too low for your *pin* photodiode detector. What alternatives are there to increase receiver sensitivity?

Answer: An optical preamplifier could increase signal strength reaching the receiver. An avalanche photodiode is more sensitive than a pin *photodiode so it should make a more sensitive receiver. Normally an APD is much less expensive.*

2. An input signal is -30 dBm (1 μW). If its data rate is 1 Gbit/s, how much energy does each pulse contain, remembering that a power of one watt equals one joule per second? If the signal is at 1.5 μm, how many photons does that correspond to, using the equation

$$E = \frac{1.989 \times 10^{-19}\, Joules/\mu m}{\lambda(\mu m)}\quad ?$$

Answer: Assuming each bit is coded so the pulse lasts exactly one billionth of a second, each pulse contains $10^{-6} \times 10^{-9}$ joule or 10^{-15} J. At a wavelength of 1.5 μm, each photon has energy of 1.326×10^{-19} J, so it takes 7541 photons to deliver that energy.

3. If a detector has a response of 1 μA/μW and the input is -30 dBm, what is the output current? How many electrons per second does this correspond to, recalling that 1 A=6.24×10^{18} electron charges per second? How many electrons would be contained in a 1-nanosecond pulse?

Answer: The output current is 1 μA, or 6.24×10^{12} electrons per second. A 1-ns pulse contains 6240 electrons.

4. A silicon *pin* detector has peak sensitivity of 0.7 A/W at 800 nm. If the input signal is -20 dBm, how many electrons does a 1-ns pulse produce? Use the equations and conversion factors from questions 2 and 3.

Answer: The current output is 7 μA, or 43,680 electrons.

5. The rise time of an InGaAs *pin* photodiode is 0.005 ns. The rise time of an InGaAs avalanche photodiode at the same wavelength is 0.1 ns. All other things being equal, how much larger is the bandwidth of the *pin* photodiode? If the rise time is all that limits the bandwidth of the two devices, what are their bandwidths?

Answer: The pin *detector has 20 times higher bandwidth. Neglecting all other factors, the* pin *detector has bandwidth of 70 GHz, and the APD has bandwidth of 3.5 GHz.*

6. The breakdown voltage of a silicon APD is 100 volts. What voltage should it be operated at to have an electron multiplication factor of 20?

Answer: 90%, from Figure 11.8.

7. The dark current in a germanium *pin* photodiode is 100 nA, compared to 1 nA in an InGaAs *pin* photodiode. Suppose that all other things are equal, and the germanium detector has a sensitivity of -15 dBm, limited by dark current. What would be the sensitivity of the InGaAs detector?

Answer: -35 dBm, reflecting a factor of 100 lower dark current.

Answers to Quiz for Chapter 11

1. d

2. d

3. c

4. a

5. b.

6. b.

7. b.

8. c

9. a

10. b

11. d.

12. a.

13. c.

14. d.

<u>Supplemental Materials</u>

Bring in samples of different types of receivers, open the cases, and explain their functions. If you don't have hardware available, try using open-case photos.

Show charts illustrating spectral response of different detectors.

Find or build a receiver in which you can access the electronic signal at several points and demonstrate it in the lap, showing waveforms at various stages on an oscilloscope to show how the signals are processed.

Chapter 12:

Amplification, Regeneration,

and Wavelength Conversion

The development of optical amplifiers tremendously enhanced the capabilities of fiber-optic systems. Repeaters and regenerators serve similar functions, so I group the three together. This chapter explains how they extend transmission distance, and compares and contrasts their functions as system elements, drawing on earlier descriptions of light sources, transmitters, and receivers. Wavelength conversion is a separate function, but fits logically because it may be performed within regenerators.

<u>Overview</u>

This chapter opens with a description of the need for amplification and regeneration, and an explanation of the differences between the two processes. I then classify the functions and devices, starting with the electro-optic repeaters used in early fiber-optic systems. I then introduce the optical amplifier, and explain its advantages as an all-optical device that can simultaneously amplify many optical channels transmitted through the same fiber. Then I describe the process of signal regeneration and why it is normally done electronically rather than optically. This section also explains 2-R and 3-R regeneration (reamplification, reshaping, and retiming). To connect wavelength conversion to this discussion, I describe OEO transponders as repeaters which take input at one wavelength and generate output at a second wavelength.

After this introduction, I describe how amplifiers may be used in-line, at the transmitter end, at the receiver end, or to offset component losses. Likewise, I explain the uses of repeaters and regenerators.

The next section covers the operation and properties of optical amplifiers in general. It explains concepts of gain, saturation, the range of amplified wavelengths, and the variation of gain with wavelength. It then introduces the various types available. Table 12.1 shows which types are used in which transmission bands.

The longest section of the chapters is devoted to the dominant type, erbium-doped fiber amplifiers. This section reviews material introduced in other chapters to give a coherent picture of erbium amplifiers, showing how the operating wavelengths depend on host material and on amplifier design. Figure 12.10 shows the variation of absorption and gain across the erbium band, and gives gain profiles of C- and L-band amplifiers. The section also shows the importance of WDM and explains noise and amplified spontaneous emission. A short section mentions the existence of other doped fiber amplifiers.

Raman fiber amplifiers are becoming important for cutting-edge systems, so the following section describes Raman gain, and how it can be used for optical amplification. It also described the use of hybrid Raman amplifiers to equalize gain across a wider range of the spectrum than is possible with erbium fiber amplifiers alone.

Semiconductor optical amplifiers have been around for many years, but have been eclipsed by erbium-fiber amplifiers for general-purpose amplification. I describe their characteristics and limitations, then cover potential applications of semiconductor optical amplifiers in integrated amplifiers and for switching, modulation and other signal control.

The closing sections describe research on optical regeneration and wavelength-conversion techniques in use and in development.

Questions to Think About for Chapter 12

1. AN ERBIUM-FIBER amplifier has small-signal gain of 30 dB. If it is operated in that high-gain mode, how far can signals travel between amplifiers if fiber loss is 0.25 dB/km? Neglect all other losses.

Answer: 120 km.

2. An erbium-fiber amplifier is operated with higher total input power, so its gain is only 12 dB. How far can it transmit signals between amplifiers in the same type of fiber? Neglect all other losses.

Answer: 48 km.

3. Signals require regeneration after passing through five of the high-gain amplifiers in Question 1, but not until they have passed through 100 of the low-gain amplifiers in Question 2. What are the total spans between repeaters for the two systems.

Answer: Remember that a system with n amplifiers has n+1 spans. 120 km × 6 spans = 720 km. 101 spans × 48 km = 4848 km.

4. An erbium-fiber amplifier can generate a maximum all-line output of 20 dBm. The input on each of 40 optical channels in its operating range is -20 dBm. If the maximum all-line output can be divided equally among all channels (an unrealistically optimistic assumption), what is the highest possible gain?

Answer: Total output is 20 dBm or 100 mW. That comes to 2.5 mW per optical channel. The input power of -20 dBm is 0.01 mW, so gain is 24 dB per channel. Note this is a maximum theoretical possibility.

5. Gain in an C-band erbium-fiber amplifier varies 4 dB across the range from 1530 to 1565 nm. If you don't use any equalization, and the output power after passing through a series of amplifiers can vary no more than 25 dB, what is the longest series of amplifiers you can use? Assume the variation is the same for each amplifier in the series.

Answer: 4 dB × 6 amplifiers =24 dB, so the maximum number of amplifiers is 6. The actual variation in amplification would vary more along the series of amplifiers.

6. Equalization reduces the range of gain in a C-band erbium-fiber amplifier to 0.5 dB. Making the same assumptions, how many amplifiers can the signal pass through?

Answer: 50 amplifiers.

7. You want to install large effective area fiber in part of a system transmitting 40 optical channels to reduce nonlinear interactions between the wavelengths. Where should you install it and why?

Answer: In the sections of fiber which collect light directly from the optical amplifiers, because power is highest there. Attenuation gradually reduces power, so after some distance it will limit the amount of nonlinear effects.

Answers to Quiz for Chapter 12

1. b

2. a

3. c

4. d

5. a

6. e

7. b

8. b

9. d

10. e

11. c.

12. d.

13. a

14. b

Supplementary Material

THE SATURATION OF GAIN with increasing input power and the variation of gain across an amplifier's operating range can have considerable impact on system operation when a network configuration changes. You can build your own examples to show how the effect grows as the number of channels increases or decreases.

Developers of submarine cables worked long and hard to stretch repeater spacing, and eventually replace submerged repeaters with optical amplifiers, because of concern about reliability of complex repeaters. To make the point you can show the likelihood you can avoid failure in a repeater that contains 100 components, each of which has a 1% chance of failure in the cable's 25-year lifetime, is 0.99^{100} (=0.366).

Ask your students to consider the tradeoffs of stretching repeater spacing on a submarine cable in order to improve reliability. Recall that noise levels increase with the gain per amplifier.

Look for network maps (or draw sample ones) to show where regenerators can be installed in terrestrial networks.

Chapter 13:

Connectors and Splices

$\mathbf{M}$aking reliable low-loss connectors and splices were key issues in the early days of fiber-optic communications. Connectors and splices both transfer signals between input and output fibers, but do so in different ways. This chapter covers their applications, the basic concepts of light transfer into and out of fibers, and practical connectors and splices.

Overview

Connectors and splices make both optical and mechanical connections between a pair of fibers. Connectors may also be used to connect a fiber with a transmitter, receiver, or other component. The essential difference between the two is that connectors are designed to be plugged and unplugged repeatedly, so network configurations can be changed. Splices are intended to be permanent junctions. The implications of this difference are also described. The introduction draws analogies with electronic connections and splices to help students familiar with electronics.

The next section describes the optical effects behind fiber-to-fiber attenuation, most of which apply to both splices and connectors. They include features inherent to the fibers, as well as features arising from fiber mounting and connector installation. They include overlap of fiber cores, alignment of fiber axes, numerical aperture, spacing between fibers, end reflections, and other end losses. A short section describes other impacts of internal reflections on system performance.

After explaining the general optical principles shared by connectors and splices, the chapter turns to connectors, starting with the mechanical features that make them rematable. This includes size, durability, and ability to withstand environmental conditions. After describing the structural elements of connectors in general, I mention connector installation procedures, without going into detail. Then I describe major types of connectors, with illustrations to show their operation. This material includes both traditional SC, ST, and FC connectors, and the newer small form factor connectors, including the LC and duplex MT-RJ.

The balance of the chapter is devoted to splicing, with an introduction stressing the differences between splices and connectors. A description of optical and mechanical considerations of splices follows. Then I describe and compare fusion and mechanical splices, with schematic illustrations of both. Splice enclosures also are shown.

This is a descriptive chapter, not one on connector and splice installation techniques. The goal is to introduce students to the concepts of splices and connectors, not to detail field or factory procedures. However, this is also the sort of material that would benefit from hands-on opportunities, particularly to compare different types of connectors. Your explanations would benefit from a few minutes spent playing with each type, even if you don't have enough samples for the entire class.

Questions to Think About for Chapter 13

1. What is the loss caused by core-diameter mismatch when going from a single-mode step-index fiber with 9-μm core to a graded-index multimode fiber with 50-μm core?

Answer: No extra loss because the graded-index core collects all light from the smaller single-mode fiber.

2. You are transferring light from a with 62.5/125-μm graded-index fiber with NA=0.275 to a single-mode step-index fiber with 9-μm core and NA=0.13. You're going to lose a lot of light from the core-size mismatch. How much loss comes from the NA mismatch? How much from the area mismatch, assuming even light distribution?

Answer: Recall that the formula for NA mismatch loss is:

$$\mathrm{Loss}(dB) = 10\,\log_{10}\left(\frac{NA_2}{NA_1}\right)^2 = 10\,\log_{10}\left(\frac{0.13}{0.275}\right)^2$$

That factor gives you a loss of 6.5 dB. Loss from area mismatch is the ratio of the areas, or 17 dB.

3. You saw earlier that Fresnel reflection loss for an air gap between a pair of fiber is 0.32 dB. Recall that the loss depends on the difference between the refractive indexes of the material in the gap and the glass. If you have water with refractive index of 1.33 in the gap, what is the Fresnel loss.

Answer: The formula for fresnel loss at one fiber end is

$$R = \left(\frac{n_{fiber} - n}{n_{fiber} + n}\right)^2 = \left(\frac{1.47 - 1.33}{1.47 + 1.33}\right)^2 = 0.0025$$

or 0.25% per surface, which corresponds to 0.01 dB. For two surfaces the loss is 0.02 dB. Water is not a good index-matching fluid, but the calculation shows the index match does not have to be perfect to limit loss. It may need to be better to prevent back-reflection.

4. If two step-index fibers with core radius *a* are offset a distance *d* from each other, as shown in Figure 13.1, the area of the two cores that overlap is

$$A_{overlap} = 2a^2 \arccos\left(\frac{d}{2a}\right) - d\left(a^2 - \frac{d^2}{4}\right)^{0.5}$$

Suppose your connector makes a 1-μm error in aligning the otherwise identical cores of two step-index single-mode fibers with 9-μm cores. How much loss does that cause?

Answer: Plug the numbers into the formula, and you find the area of overlap is 54.6 square micrometers, compared to a core area of 63.6 square micrometers, or a fractional overlap of 85.9%, corresponding to a loss of 0.66 dB.

5. Use the formula of Question 4, go back and estimate how precisely the same step-index single-mode fibers have to be aligned to have offset loss of only 0.3 dB. (Hint: You can try the formula for different values of offset if you program it into a computer spreadsheet.)

Answer: Successive approximation in a spreadsheet gives 0.472 μm.

6. Why can't two twist-on connectors be assembled into a unit as a duplex connector?

Answer: Because they have to be twisted in place, which inevitably rotates one fiber relative to the other.

7. A major telephone carrier puts you in charge of field repairs for a major urban center. You need to outfit a special truck for skilled technicians to use in repairing breaks in overhead and buried cables. What type of splicer do you buy?

Answer: A fusion splicer, for its lower loss and because it's likely to see heavy use.

8. A large retail company hires you to manage its data-transmission networks. The company has one office with fiber running to a dozen desks in each state. You want to supply every office with a repair kit in case someone trips over a cable. What type of splice equipment do you buy?

Answer: Mechanical splices, which are easier for untrained people to install. You doubt anyone will need it, but you don't want to have to fly someone in to do repairs.

Answers to Quiz for Chapter 13

1. b.

2. c.

3. d.

4. c.

5. b.

6. c.

7. a.

8. b.

9. a.

10. b.

11. c.

12. b.

13. d.

14. a.

Supplementary Material

An ideal supplement for the connector discussion is a set of terminated cables to pass around, so students can examine them.

Company catalogs and web sites are a useful resource. Some include product literature that shows how connectors work. Figure 13.14c, which shows four stages in the mating of the VF-45 connector, is a good example.

Catalogs and web sites also can provide photos of splices and splicing components.

Bob Chomycz, *Fiber Optic Installer's Field Manual*, (McGraw Hill, 2000), illustrates some installation procedures.

You may be able to get videos showing procedures for installing splices and connectors from manufacturers.

If you have access to a fusion splicer, you can demonstrate it for your class, or allow students to make their own splices. It's easy with a little coaching; I've done it at trade shows. You might be able to talk a friendly sales rep or local technician into a demonstration for your class. Fusion splicers are well-designed machines, and a few minutes showing how one is designed to meet demanding requirements would help students understand precision equipment as well as fiber splicing. Manufacturers' catalogs are good sources of illustrations showing the workings of mechanical splices.

If your school has a fiber network, someone in network maintenance is likely to have mechanical splicing equipment for emergency repairs.

Chapter 14:

Couplers and Other Passive Components

This chapter covers couplers as well as other types of 'passive' components. Passive components in this sense are those that require no outside power. Wavelength-selective optics are covered separately in Chapter 15. Active components, such as switches and modulators, which require additional power or operate on the signal, are covered in Chapter 16. This grouping is somewhat arbitrary, but it divides a large and important range of components into three manageable chapters.

<u>Overview</u>

Couplers are the most important passive optical components, and the bulk of this chapter is devoted to them. The introduction concentrates on how optical couplers divide signals, and how they differ from electronic couplers. The key point is that optical signals are in the form of photons, which have to go one way or the other through an optical coupler with two outputs. If the two outputs are equal, each one gets only half the photons. Table 14.1 shows how this works for couplers with more output ports. In contrast, if an electrical coupler is operating with an ideal voltage source and delivering signals to a high impedance, both outputs see the same voltage. It's worth stressing this difference if many of your students are familiar with electronics.

The next section covers important operational characteristics of couplers, including the number of input and outputs, how signals are split, and the degree of directionality. Many couplers are sensitive to wavelength, and this section gives some examples, but the main use of wavelength-selective couplers is in WDM, covered in Chapter 15.

The following section covers the types of couplers, and the technologies used for coupling, It explains the common features of bulk and micro optics, using a beamsplitter as an example. A partially transparent mirror is a good classroom prop. Then I explain the optics of fused-fiber couplers and planar waveguide couplers. Note the that light splitting differs between single-mode and multi-mode couplers.

The confusing concept of "active" couplers is briefly explained. This term is sometimes used for electro-optic repeaters in which one receiver drives two transmitters. This approach has practical advantages in local-area networks, but it's really a special-purpose repeater rather than the sort of coupler covered in the rest of this chapter.

The rest of the chapter describes three other important types of passive components. The section on attenuators describes how they limit light intensity, and distinguishes them from wavelength-selective optical filters. Students may need to have the difference from resistors explained. Optical isolators are important because they block back-reflected light. This section gives only a single example, a polarization-dependent optical isolator design in Figure 14.11 that is easy for students to understand.

Like optical isolators, optical circulators restrict the direction of light travel. You can think of them as optical one-way streets, or the optical counterpart of a traffic rotary, which allow light to travel in one direction around a loop but not the other. Figure 14.12 shows how complicated the designs can get. As with optical isolators, other designs are possible.

<u>Questions to Think About for Chapter 14</u>

1. Suppose your input signal is -10 dBm and your receivers require a signal of at least 0 dBµ. You want to distribute signals to as many terminals as possible. If there is 3 dB of fiber loss between you and each receiver, how many terminals can you deliver signals to? How much coupler loss does this correspond to on each channel. Assume you can buy a star coupler with as many ports as you want which has no excess loss.

Answer: The coupler output has to be at least 3 dB higher than 0 dBµ, or +3 dBµ. It's easiest to consider this problem in watts, so convert that to 2 µW. The input power is 0.1 mW or 100 µW, 50 times higher. If there is no excess loss, you can distribute signals to 50 terminals. This corresponds to 17 dB loss.

2. Suppose that all star couplers available for the system described in Question 1 have excess loss of 3 dB. How many terminals can you reach with these couplers, and what is the total loss per channel?

Answer: The excess loss effectively reduces the input signal to -13 dBm, or 50 µW. If you need to distribute 2 µW to each output, you can serve 25 terminals. The loss per channel is still 17 dB, because the input and output powers are the same, but it is the sum of 14 dB of loss in dividing the signal and 3 dB of excess loss. In reality, the excess loss would depend on the number of terminals.

3. An alternative design is to cascade a series of 3-dB T couplers. The first splits the signal in half, then each output has its own 3-dB coupler, dividing that output in half, yielding one-quarter of the original output. Adding more layers further divides the signal. Suppose you can get as many 3-dB couplers as you want and each one has no excess loss. How many terminals can you divide signals among in Question 1?

Answer: You need at least 3 dBµ output from each terminal of the coupler cascade, so the output loss per channel can be no more than 17 dB. Each T coupler has 3-dB loss, so you can have no more than 5 in series, for a loss of 15 dB/channel. The number of outputs from five-layer cascade is 2^5 or 32. (More generally, if the number of layers is N, the number of outputs is 2^N.) Note that this design includes 2 dB of margin per channel; students will learn the importance of system margin later.

4. A local area network includes 90-10 couplers, which split 10% of the input signal and deliver it to a local terminal. Suppose you have 10 of them in series and the input power is -10 dBm. What is the power delivered by the last coupler out each of its ports.

Answer: The loss of the coupler is 10 dB in the 10% direction and 0.46 dB out the 90% port. The total loss is 9x0.46 dB=4.14 dB (the signal emerges from the 90% direction in the first 9 couplers), plus 10 dB (for the final coupler where it emerges from the 10% port, or 14.14 dB. This leaves an output of -14 dBm, or 4 µW.

5. If your receiver requires 1 µW of power, how many more 90-10 couplers could you have in series before the 10% side does not deliver enough power for reliable operation. Assume the same -10 dBm input as in Question 4.

Answer: The total allowable loss is 20 dB. Of this, 10 dB comes in the final coupler, where the 10% channel receives a signal 10 dB below the input to that final stage. That means that a total of 10 dB of loss can occur in the other N couplers. Divide 10 dB by 0.46 dB/coupler, and you have enough signal to pass through 21 couplers before reaching the final one. That means the total number is 22 couplers.

6. An optical amplifier delivers 0.5 mW/channel on each of 32 channels. You want to monitor its performance by diverting a small portion of its output to an optical performance monitor which requires 1 dBµ input per optical channel. What fraction of the output power do you need to divert to the performance monitor.

Answer: At least 1 part in 500.

7. Neglecting excess internal losses, what is the difference in attenuation between the following two optical circulators. The first uses the simple optical isolators of figure 14.10, one oriented from Port 1 to 2, and the second from Port 2 to 3, with a 50-50 T coupler splitting input signals from Port 2 between two routes. The other is the more complex optical circulator of Figure 14.11.

Answer: Each optical isolator has 3-dB loss because it loses one polarization. This means 3 dB loss from Port 1 to Port 2. Port 2 also has 3-dB entrance loss because signals are split in two directions, so loss from Port 2 to Port 3 is 6 dB, 3 dB of coupler loss plus 3 dB of isolator loss. The optical circulator of Figure 14.11 maintains both polarizations, so it will have no 3-dB losses (but in practice it will have some loss).

Answers to Quiz for Chapter 14

1. c.

2. e.

3. c.

4. b.

5. a.

6. c.

7. b.

8. e.

9. c.

10. c.

Solved Quiz Problem

Problem 2.

In this example, you need to calculate total output power by summing all the output signals, so you must convert decibel units to milliwatts. Start by converting –10 dBm to 0.1 mW and –30 dBm to 0.001 mW.

$$Excess\ loss = -10\log\left(\frac{\sum P_{out}}{P_{in}}\right) = -10\log\left(\frac{20 \times 0.001\ mW}{0.1\ mW}\right) = -10\log(0.2) = 7dB$$

Supplementary Material

A beamsplitter or partly transparent mirror is a simple visual example of a coupler, if you want to show the idea to a whole class.

This use of polarizers in the optical isolator shown in Figure 14.11 causes it to have loss of 3 dB per pass because the design throws away one polarization. You can find alternative designs that are not polarization-sensitive at **http://www.fdk.co.jp/laboratory/hikariai-e.html**. See if you can get students to design ones that are polarization insensitive. (You can drop the hint of using alternative paths for the two polarizations.)

The optical circulator shown in Figure 14.12 has a quite complex optical path. A good exercise for students is to trace the path through carefully to verify that it performs its functions. A tricky exercise is to change the signs or arrows in a couple of places and challenge students to spot the mistake. I must admit that one incorrect sign slipped past me and got into the version of the figure (14.11) in the fourth edition. If you have a copy of that edition, make photocopies and challenge students to trace the path and find the typo.

Chapter 15:

───

Wavelength-Division Multiplexing Optics

W DM optics are complex and important, and are treated in a chapter of their own because they depend on different principles than other types of passive optics. This chapter mentions potential capacity, but notes that after the bubble much of this potential capacity remains unused. Although I avoid most acronyms in the text, WDM is so important (and the full term so long) that I make an exception for it.

Overview

The chapter opens with a description of the operations which may be performed in WDM systems. Then it shifts to the design of WDM systems themselves. The two main families of WDM systems differ primarily in total capacity and design goals. Dense WDM (DWDM), originally developed for long-distance transmission, squeezes many optical channels tightly together to transmit as high a total data rate as possible. Coarse WDM (CWDM), developed to reduce costs, does not try to pack signals together as densely. Figure 15.2 contrasts their channel spacing to scale. This section also shows the importance of channel spacing, and describes related concepts including add-drop multiplexers, wavelength routing, channel equalization, and advanced optical networking.

Optical filters are widely used to select wavelength, so the next section covers the important technology of thin-film interference filters. Interference filters can selectively transmit very narrow slices of the spectrum; they also can be designed to pass wider bands of wavelengths, or transmit only wavelengths shorter or longer than a certain cutoff value. Equalizing filters selectively attenuate certain wavelengths to compensate for amplifiers with stronger gain at those wavelengths, to give a uniform net gain across a wider range of wavelengths. Most filters are fixed in their transmission characteristics, but tunable filters offer important advantages, so I cover a couple of approaches. Students who have taken an optics course may have been exposed to the basic concepts, but would not necessarily know the fiber-optic portions of this section.

The next section compares how wavelength-selection technologies are used in WDM systems. It covers interference filters, fiber Bragg gratings, fused fiber couplers, Mach-Zehnder interferometers or interleavers, bulk diffraction gratings, and waveguide arrays. The first three technologies were introduced earlier, so this section concentrates on how they are used for WDM systems. You can refer students to Chapter 7 for fiber Bragg gratings and to Chapter 14 for fused couplers. Diagrams show how multiple components are assembled to combine or separate wavelengths in a WDM system. The illustrations show demultiplexing; inputs can be combined by reversing the process.

Mach-Zehnder interferometers are similar in concept to fused fiber couplers, but unlike fused fiber couplers are designed to separate closely spaced wavelengths. They function as interleavers, shuffling together or unshuffling optical channels spaced at uniform frequency intervals. (Although figures conventionally show wavelength, the spacing actually is based on frequency.)

Bulk diffraction gratings were mentioned earlier; the section on them describes how they are used for WDM. Similar devices are used in instrumentation.

54

The description of waveguide arrays simplifies their operation to make them comprehensible to students entering the field. Their operation depends on differences in the lengths of the curved guides. Some students may miss this distinction because all the guides are drawn parallel to each other, but following the lines shows they differ in length. The input signal is split among the waveguides in the array; the output of the waveguides interferes in the output coupler, which directs wavelengths to different ports.

The final section points out that different technologies may be combined to perform different functions in a WDM system.

Note that although the optical systems may accommodate large numbers of optical channels, all of those are not likely to be populated in real-world systems.

<u>Questions to Think About for Chapter 15</u>

1. AN ERBIUM-DOPED fiber amplifier can transmit signals at wavelengths between 1530 and 1565 nm. How many optical channels can you fit in this range with 200 GHz spacing? How many channels with 100 GHz spacing?

Answer: First calculate the frequencies corresponding to the wavelengths (it's best to use the exact value of the speed of light, 2.997952×10^8 m/second.) 1530 nm corresponds to a frequency of 195.943 THz. 1565 nm corresponds to 191.560 THz. The difference is 4.383 THz. Dividing by 200 GHz leaves 21.9 200 GHz slots. Dividing by 100 GHz leaves 43.89 100 GHz slots.

2. A transatlantic fiber-optic cable contains 100 optical amplifiers. It needs equalizing filters to balance the gain of the erbium-fiber amplifiers across their operating ranges. If the receivers used on the system have a dynamic range of 20 dB, how closely do the equalizing filters have to balance gain. (Assume all filters and amplifiers are identical.)

Answer: Within 0.2 dB across the spectrum. Note that this problem is an oversimplification of the real design constraints. More powerful channels tend to drain power from weaker channels as the difference builds up.

3. A typical interference filter for demultiplexing 100-GHz channels has 0.5 dB loss on the reflected channels and 2.0 dB loss on the transmitted channels. How much loss does the signal suffer on the first channel picked off (λ_1) in Figure 15.7? What is the loss for the last channel of eight channels picked off (λ_8) in a similar arrangement? What channel in an 8-channel system suffers the highest total loss?

Answer: The first channel has 2-dB loss from transmission but no loss for reflection, a total loss of 2 dB. The eighth channel is reflected 7 times, so it suffers a total of 3.5 dB loss. The highest loss is on the seventh channel, reflected 6 times (3 dB total) and transmitted once (2 dB), for a total of 5 dB loss.

4. A typical fiber Bragg grating has 99.9% reflection (0.0043 dB loss) and 0.2 dB loss for transmitted wavelengths. Assume loss of 1 dB in the optical circulator. What is the loss for the first channel of 8 channels picked off in a cascaded series of fiber Bragg gratings? What are losses for the 7th and 8th channels?

Answer: The first channel suffers 1 dB loss in the circulator and 0.0043 dB loss from reflection, a total of 1.0043 dB. The seventh channel is transmitted 6 times (1.2 dB total) and reflected and passed through the optical circulator once (1.0043 dB) for a total loss of 2.2 dB. The eighth channel is transmitted 7 times, for a loss of 1.4 dB, but does not have to go through the optical circulator.

5. What should the difference in path lengths be in a Mach-Zehnder interferometer designed to interleave optical channels separated by 50 GHz? By 200 GHz? Assume the refractive index of the material is 1.5 and is uniform for both arms.

Answer: Approximating with c = 3×10⁸ m/sec, and using the formula

$$\Delta L = \frac{c}{2n\Delta v}$$

for 50 GHz the difference is 2 mm. For 200 GHz the difference is 0.5 mm.

6. You want to separate 16 optical channels that are uniformly spaced 50 GHz apart with optical interleavers. How many interleavers do you need? How many interleavers does each optical channel pass through?

Answer: 16 optical channels require a total of 15 interleavers. The number required to separate N channels generally is N-1. Each channel would pass through 4 interleavers in series. You can see this by adding a level to Figure 15.12.

7. A 40-channel arrayed waveguide demultiplexer has average loss of 8 dB for each channel processed. How does this compare to the highest loss of the 40-channel interference-filter demultiplexer shown in Figure 15.8. You can use the results from question 3 to give you the loss for the 8-channel demultiplexing boxes. What's the minimum loss?

Answer: Channels 33-40 are transmitted twice and reflected once outside the 8-channel demultiplexers, for an extra loss of 4 dB in transmission and 0.5 dB in reflection, or a total of 4.5 dB. The highest loss in the 8-channel demultiplexer is 5 dB for channel 7, which then experiences a total loss of 9.5 dB. The minimum loss is for channels 9-16, reflected twice and not transmitted at all, for a total loss of 1.0 dB. The first channel has lowest loss of 2.0 dB, for minimum total loss of 3.0 dB.

———————————

ANSWERS TO QUIZ FOR Chapter 15

1. b.

2. c.

3. a.

4. b.

5. d.

6. b.

7. c.

8. b.

9. b.

10. d.

11. c.

12. c.

<u>Supplementary Material</u>

Color illustrations showing optical channels in different colors can help you explain the operation of a WDM system. Some are included in the PowerPoint presentations on the CD-ROM.

An interference filter for visible wavelengths is an excellent way to show the operation of interference filters. You can sometimes get samples or rejected filters from manufacturers. A large one works best to show to a class. Show how the filter transmits one wavelength but reflects other visible wavelengths. You can contrast this with a simple photographic color filter, which transmits some colors and absorbs other wavelengths.

Chapter 16:

Optical Switches, Modulators, and Other Active Components

Active components are those which act upon light signals, and generally require some outside power. External modulators and optical switches are the most important active components in current systems, and optical switches are vital elements in optical networks. This chapter describes their roles in fiber-optic systems and the technologies used, as well as looking at the emerging technologies of wavelength conversion and integrated optics.

Overview

I open the chapter by defining active components and giving some examples. Although the concept of "active" components is important, it is not worth spending huge amounts of time trying to draw the finest possible lines between active and passive components.

Modulators are a logical starting point because their functions are related to switches and in some senses more general. A modulator adjusts output power over a continuous range, from all on to all off. Although most modulators are digital, the technology can be used for analog modulation, which continuously varies intensity. You can explain a modulator as a sort of shutter which adjusts its transparency between 0% and 100%, varying the intensity of the transmitted light. In the electro-optic waveguide modulator, the transparency is varied by causing a phase shift between light in two parallel waveguides, with interference effects modulating intensity. Electro-absorption modulators are diodes modulated with a reverse-biased signal, so they transmit light when the bias is off and absorb it when the bias is on. Modulators have to operate on extremely fast time scales, smaller then the bit rate, but I note that variable optical attenuators perform the same function on a longer time scale.

The next section describes the varied switching requirements in telecommunication systems. I describe protection switching, provisioning, reconfiguration, cross connects, switching fabrics, circuit switching, wavelength routing, and routing. I also describe the use of optical switching to select individual wavelengths on a fiber carrying a WDM signal. A subsection describes the differences between circuit switching and routing. That may seem peripheral to fiber-optic systems, but in my experience confusion between circuit and packet switching can be a major conceptual stumbling block for students trying to understand telecommunications. I also explain the differences between transparent and opaque switches.

The following section describes the switching technologies that are most widely used today, including how they are used. Opto-mechanical switches are the simplest to understand; they are the optical counterparts of mechanical switches used to turn the lights on and off. MEMS technology can take different forms; it was hot during the bubble, but interest has cooled. The ingenious mechanism of the bubble switch is worth describing carefully. Electro-optic switches are a standard textbook example, based on electro-optic modulators, but are not widely used in practical systems. Thermo-optic switches rely on the same interference mechanism to switch outputs, with the change induced by temperature changes. Liquid-crystal switches also are important, and other types are emerging.

The final few pages cover emerging technologies. WDM optical networks will require wavelength converters to manage signals; you can compare not being able to change wavelengths to not being able to shift lanes on a freeway. The concept of integrated optics has been around for decades, but is only finding practical applications now.

Questions to Think About for Chapter 16

1. An external modulator with a 20-dB extinction ratio modulates the output of a 1-mW laser. The signal then passes through 20 km of fiber with loss of 0.5 dB/km. Neglecting other losses, what are the powers at the detector when the light is off and when it is on?

Answer: Power is 0.1 mW when the beam is on and 0.001 mW when the beam is off.

2. What are the power levels in Question 1 if the external modulator has insertion loss of 3 dB?

Answer: Power is 0.05 mW when the beam is on and 0.0005 mW when the beam is 'off'.

3. What are the power levels for Question 1 if the external modulator has 3-dB insertion loss and an extinction ratio of 10 dB?

Answer: Power is 0.05 mW when the beam is on and 0.005 mW when the beam is off.

4. One important issue in switch design is the number of elements required to switch the Suppose you have a simple cross-connect switch such as the one shown in Figure 16.4, with a single switch element at each node which either transmits or reflects the beam. If you have 10 inputs and 10 outputs, how many switching elements do you need? What if you have 100 inputs and 100 outputs?

Answers: If the number of inputs and outputs are the same, the number of switching elements scales as the square of the number of ports N. Thus a 10x10 switch requires 100 elements and a 100×100 switch requires 10,000 elements. This is a significant limit on some cross-connect designs.

5. A tilting-mirror switch can reflect the light input from a single input port to any of N output ports. With this design, how many switching elements do you need for a 10×10 switch? A 100×100 switch? Assume the tilting mirror can point the beam at as many output ports as needed.

Answers: This design scales as the number of ports N. Thus a 10×10 switch requires 10 elements and a 100×100 switch requires 100 elements. The trick here is being able to point the tilting mirror reliably at each of 100 possible output ports. Actual designs are more complex.

6. How does the bubble switch shown in Figure 16.9 scale? What number of bubble-waveguide intersections do you need for an $N \times N$ optical cross-connect?

Answer: N^2 intersections. The switch can either reflect or transmit light on its path, like the design of Question 4.

Answers to Quiz for Chapter 16

1. d.

2. a.

3. b.

4. c.

5. b.

6. a.

7. c.

8. b.

9. b.

10. c.

11. d.

12. c.

13. d.

14. e

<u>Supplementary Material</u>

Note that the presentation on the CD-ROM contains color versions of a few figures.

You can illustrate the idea of an opto-mechanical switch by bending a fiber back and forth. Large-core plastic fibers show up best.

You can simulate a tilting-mirror MEMS switch with a hand-held flat mirror and a fixed laser pointer, scanning the laser spot across the ceiling or a wall.

Jeff Hecht, "All-Optical Switches," *Laser Focus World* August 2001

Jeff Hecht, "Wavelength Conversion," *Laser Focus World*, April 2001

Chapter 17:

Fiber-Optic Measurements

Fiber-optic measurements use techniques from both optics and communications. This chapter introduces both types of measurements, but puts particular emphasis on explaining optical measurements and their peculiarities. It may require extra time because it introduces some terminology and ideas peculiar to optics. This chapter emphasizes concepts, so it does not give detailed procedures. Chapter 18 covers test equipment and troubleshooting, and you should cover the two chapters in sequence.

Overview

Fiber-optic measurements require some background in optics, electronics and communications. I emphasize the optical side, because few students have had much exposure to optical measurements. The goal is to make students familiar with fiber measurement concepts, but not to make them skilled measurement specialists. That would bee a course in itself.

The chapter starts with an introduction to the basics of optical power measurement, which describes optical quantities and their meanings, summarized in Table 17.1. Simple equations show basic relationships. I compare optical and electronic power measurements, for students familiar with electronics. I also explain peak and average power, and quirks of optical power measurement that. A subsection on the types of power measurement describes the role of detectors. It also distinguishes two pairs of often-confused terms: irradiance and intensity, and radiometry and photometry.

The section on wavelength measurement stresses the need for precise calculations in certain cases. One example shows the difference when calculations are made with the speed of light rounded to 300,000 kilometers per second, and those made with the speed of light given to seven significant digits. Students may be surprised that the difference is enough to change the wavelength by more than 1 nm, more than one 100-GHz optical channel spacing. Then the section describes relative and absolute measurements of wavelength, measurements of spectral linewidth, and spectral response.

The next sections cover phase and interference measurements and polarization measurements. Both are important to understand. The operation of many advanced devices such as optical switches depend on phase and interference effects. Polarization-dependent loss and gain and polarization-mode dispersion can pose serious problems in systems.

A section on time and bandwidth measurements covers pulse timing, rise and fall time, repetition rate, and pulse spacing, with Figure 17.7 showing what the terms mean on a sample wave form. I also describe frequency response measurements. The section on signal quality describes signal to noise ratio, bit error rate, and eye pattern analysis.

A final section covers fiber-specific measurements, including measurement standards for optical fibers, and standard measurement assumptions. A table lists standard organizations, but the details of specific standards are beyond the scope of this book. The rest of this section briefly describes common measurements such as attenuation, fiber continuity, mode-field diameter, numerical aperture, and cutoff wavelength.

Questions to Think About for Chapter 17

1. A 1-Gbit/s signal has an average power of 1 μW at the receiver. What is the average energy in each pulse?

Answer: Divide the average power in watts by the number of pulses, to get average pulse power in joules—10^{-15}.

2. A 10-Gbit/s signal has an average power of 1 μW at the receiver. What is the average energy in each pulse?

Answer: Divide the average power in watts by the number of pulses, to get average pulse power in joules—10^{-16}.

3. The wavelength being transmitted in examples 1 and 2 is 1550 nm. How many photons does each pulse contain?

Answer: Calculate the energy per 1550 nm photon using the formula $E=hc/\lambda$ recalling that h is 6.63×10^{-34} joule-second. This gives: 1.28×10^{-19} joule/photon.

Then divide the average power per pulse by the number of photons. For 1 Gbit/s, the average pulse contains 7793 photons. For 10 Gbit/s, the average pulse contains 10 times fewer, or 779. That difference is very important in receiver performance.

4. Suppose these systems were operating at a wavelength of 850 nm. How many photons would be in each pulse at average power of 1 μW and data rates of 1 and 10 Gbit/s? Assuming that noise levels are constant relative to photons per bit, how would that affect signal to noise ratios?

Answer: You can repeat the calculations for Question 3 with the photon energy at 850 nm, which give 4274 photons per bit at 1 Gbit/s and 427 photons per bit at 10 Gbit/s. All other things being equal, that reduction in photon numbers would decrease signal-to-noise ratio by 2.6 dB.

5. The air pressure at the Keck Telescope on the top of Mauna Kea in Hawaii, 4.2 kilometers above sea level, is about 60% of that at sea level. The refractive index of air is 1.000273 at standard temperature and pressure. Assume that the refractive index of air is proportional to density. What are the wavelengths of light with a frequency of 193.1 THz in a vacuum, in air at sea level (standard temperature and pressure), and at the top of Mauna Kea? Be sure to use the exact value of the speed of light, 299,792.5 km/s. How does this compare with the shift in wavelength between optical channels at 50 GHz spacing at the same frequency?

Answer: Use the formula $\lambda=c/nv$ for wavelength. Calculate the refractive index for Mauna Kea by taking the difference between the refractive index of air at sea level and the index of a vacuum, multiplying by 0.6, then adding one, because the difference in index is between air and vacuum (which is one, not zero). The answers are 1552.10 nm at sea level, 1552.27 nm on Mauna Kea, and 1552.52 nm in a vacuum. The difference between sea level and a vacuum is slightly larger than the 0.4 nm channel spacing at 50 GHz. The wavelength on Mauna Kea is 0.17 nm longer than at sea level, about 40% of a 50-GHz optical channel.

---

6. WHAT IS A FREQUENCY difference of 100 GHz equivalent to at the 850 nm wavelength of gallium arsenide lasers?

Answer: Calculate the frequency corresponding to 850 nm from the speed of light and wavelength; the answer is 352.692 THz. Then add 100 GHz and calculate the wavelength, which comes to 849.76 nm, showing a shift of 0.24 nm. This is a smaller fraction of wavelength than at 1550 nm because the 100-GHz increment is a smaller change in frequency at the shorter wavelength.

<u>Answers to Quiz for Chapter 17</u>

1. b.

2. a.

3. c.

4. d.

5. b.

6. e.

7. c.

8. b.

9. e.

10. a.

11. d.

12. b.

13. e.

14. a.

15. d.

Solved Quiz Problem

Problem 9.

Over a period of one second, a 1-mW beam delivers a total of 1 mJ or energy. To calculate the number of photons in 1 mJ, first use Planck's law to calculate photon energy.

$$E = h\nu = \left(6.626 \times 10^{-34}\right) \times \left(1.931 \times 10^{14}\right) = 1.2795 \ J / photon$$

Then divide that value into the total energy (0.001 mJ) to find the number of photons needed to provide that energy.

$$\frac{0.001 \ J}{1.2795 \times 10^{-19} \ J / photon} = 7.8157 \times 10^{15} \ photons / mJ$$

If you want to show how the number of photons can limit signal reception, you can ask students to calculate how many photons beams of 1 mW or 1 μW could deliver in the 1-ns bit interval for a 1 Gbit/s signal or the 0.1 ns bit interval for a 10-Gbit/s signal.

Supplementary Material

A good advanced treatment of fiber measurement is Dennis Derickson ed, *Fiber Optic Test and Measurement* (Prentice Hall PTR, Upper Saddle River, NJ, 1998)

A fiber measurement laboratory exercise can be very valuable if you can have access to suitable instruments. Measurement theory is helpful, but nothing substitutes for hands-on experience. If you have access to a sampling of fibers and cables, an optical power meter and matching light source will be invaluable. If you have sources and detectors for different wavelengths, you can show (or students can learn firsthand) the importance of matching sources and detectors.

Catalogs from some test-equipment companies are full of useful information, particularly on the measurement instruments listed in Chapter 18. Agilent and Exfo are companies that have provided helpful catalogs. Much of this information now is available on web sites, though it may take some digging.

Chapter 18:

Troubleshooting and Test Equipment

Troubleshooting is a much-requested topic. I have grouped it with test equipment because effective troubleshooting in fiber optics requires the use of test equipment. Chapter 17 introduced measurement concepts; this chapter talks about specific measurement instruments and explains troubleshooting procedures in general. It is not a cookbook or detailed guide to specific procedures. Those are the roles of more specialized books and courses.

Overview

The chapter opens with a brief description of fiber-optic troubleshooting, pointing out the many ways that things can go wrong. I also differentiate between testing, installation, and troubleshooting, all of which require optical measurements, and note that measurements may interrupt signal transmission.

A long section covers important types of test equipment, typically describing each one in a few paragraphs. My intent is largely to define what's what and how instruments work, not to explain how to use specific instruments. I give details and sample measurements for a few important instruments. One is the optical time-domain reflectometer, where Figure 18.4 shows how to interpret measurements, and Figure 18.5 shows an actual instrument readout. Another is the optical spectrum analyzer, which provides both an analog readout of the waveform and digital readings of wavelength, power, and signal to noise ratio as shown Figure 18.7. Figure 18.8 shows the workings of an optical performance monitor.

The final section gives general troubleshooting procedures for fiber-optic systems. These are not exhaustive or entirely original; some ideas are drawn from the resources listed as further reading in the book. The idea is to show the principles involved.

Questions to Think About for Chapter 18

1. Devise a trouble-shooting procedure for testing your ability to connect to the Internet using a personal computer, an external modem, and a standard telephone line.

Answers: Details will vary, but students should check their software, their modem port, power to their modem and computer, modem operation, the wires to the modem and wall outlet, connectors on the cables, and the phone wiring in their building and between the building and the telephone switching office. They also should check their Internet Service Provider. The principles are the same for broadband connections using DSL or cable modems.

2. Which is more likely to fail: An old fiber-optic cable left undisturbed in an underground duct, or a new office cable that was plugged into a new computer?

Answer: The office cable, because it was just moved. Undisturbed cables generally don't fail on their own accord.

3. What is the advantage of measuring fiber attenuation at 1625 nm?

Answer: That wavelength is not used in WDM systems, so transmitting signals at that wavelength will not interrupt other optical channels.

4. You want to test a WDM system by transmitting signals through it one wavelength at a time at the same power generated by the standard transmitters. What sort of light source should you use.

Answer: A tunable laser.

5. You test a 50-kilometer cable with an OTDR and find a sharp peak at 25 kilometers and no signal returning from greater distances. What does this tell you about the cable?

Answer: The fiber is broken at 25 kilometers.

6. When the moving arm in a wavelength meter moves 7 millimeters, the instrument counts 9,000 fringes. What's the difference between the wavelength in vacuum and the wavelength in air (n=1.000273)?

Answer: For 7 mm travel and 9000 fringes, the wavelength in air is

$$\lambda = \frac{2L}{N} = \frac{2 \times 7000 \, \mu m}{9000} = 1.55556 \, \mu m$$

The wavelength in vacuum is the wavelength in air divided by the refractive index 1.000273, or 1.55513 μm. The differences is 0.43 nm, slightly more than 50 GHz. This is just over half the ITU channel spacing of 100 GHz.

Answers to Quiz for Chapter 18

1. c.

2. d.

3. a.

4. c.

5. b.

6. e.

7. a.

8. e.

9. c.

10. c.

11. a.

12. b.

Supplementary Materials

Catalogs from Agilent Technologies (www.agilent.com) and Exfo (www.exfo.com), and other companies have details on a wide variety of test instruments.

Bring test equipment into the classroom and show its operation. It's useful to show test results on systems that are unstable as well as those that are rock-solid.

Get screen shots from OTDRs and other test equipment being used for troubleshooting under a variety of conditions, and have students try to interpret them. An optical time-domain reflectometer is the best choice of an instrument to demonstrate if you only have time to demonstrate one.

A good advanced treatment of fiber measurement, recommended in Chapter 17, is Dennis Derickson ed, *Fiber Optic Test and Measurement* (Prentice Hall PTR, Upper Saddle River, NJ, 1998)

Chapter 19:

System and Optical Networking Concepts

Chapters 19-28 cover the broad area of fiber-optic systems. They are grouped loosely into two introductory chapters, two covering design concepts, and six covering system applications at different levels, from global networks to systems used inside vehicles. These chapters can be read by students who have not had a course in telecommunications systems, but they do not attempt to offer comprehensive coverage of telecommunications as a whole. The goal is only to explain the concepts important for understanding fiber-optic systems.

This chapter introduces system concepts in more systematic detail than Chapter 3. The goal is to introduce basic principles needed to understand fiber-optic systems, including concepts of telecommunications in general as well as optical networking. It also expands on concepts introduced from the component viewpoint in earlier chapters, including transmission formats, modulation, and multiplexing.

<u>Overview</u>

I open with a brief review of telecommunications in general, and the idea of optical networking. This brief introduction also mentions how business considerations have affected the structure of telecommunications systems. The business side of telecommunications is outside the scope of this book, but students should understand that the business aspects affect the technology. One example is the slump that hit in 2001 after the tremendous expansion of 1999 and 2000.

The next section gives an overview of telecommunication network structure. It shows how the network is made up of links between nodes, and mentions how signals are directed among nodes. It also introduces the concepts of standards and protocols, which are covered in Chapter 20.

The section that follows classifies networks according to their topology. It illustrates the concepts of point-to-point, point-to-multipoint, networked and switched transmission. Although these are generalizations, they are useful concepts for students trying to break the global telecommunications system into chunks they can more readily understand. I stress the difference between routing and switching on a system level, and differentiate among broadcasting, networking and switching.

Another section gives a system view of transmission formats. It reviews digital and analog transmission, and some concepts introduced earlier, such as amplitude modulation. It describes some specific digital coding schemes, and mentions error-correction codes without going into detail. Subsections introduce soliton transmission, frequency modulation, and phase-shift modulation.

The final section covers system transmission capacity, defined as the amount of information a system can transmit. Much of the section is devoted to multiplexing. I take pains to differentiate between the interleaving of bit steams in true time-division multiplexing, and the stacking of data packets in packet switching. This important difference is often overlooked. I review wavelength-division multiplexing, then describe different measures of transmission distance—amplifier spacing and end-to-end transmission distance. I then explain how these ideas come together in optical networks, and touch on cost and reliability as real-world considerations.

Questions to Think About for Chapter 19

1. INTERNET ROUTERS read headers on data packets, and use that information to direct the packets toward the proper destination. What type of network architecture would you expect to be used to connect routers in the Internet backbone? Why?

Answer: A mesh. Star or ring networks would direct the signals through all the routers, which would not use the Internet backbone efficiently.

2. Can you use headers to control the flow of data packets around a ring network in which signals pass through all the terminals anyway?

Answer: Yes. Instead of using routers to direct signals to particular terminals, the nodes of the ring could read only packets addressed to them.

3. An electronic time-division multiplexer generates signals at 1 Gbit/s. If it has eight inputs, what are their data rates?

Answer: Input data rates are equal, and add to give the output TDM rate, so each input is 125 Mbit/s.

4. How would an optical time-division multiplexer work?

Answer: It would take input optical pulses and interleave them optically. In practice, this would mean shortening the pulse durations so they would fit into the shorter bit intervals of the high-speed signals. The concept has been used in the laboratory, with optical delay lines and switches or modulators to shorten pulses, but it is not used in current practical systems.

5. How is wavelength-division multiplexing analogous to frequency-division multiplexing?

Answer: Both use separate input signals to modulate each carrier signal, and combine the modulated carriers for transmission. In WDM the carriers are light at separate wavelengths passing through a single fiber; in FDM, they are separate radio frequencies merged into a broadband signal or transmitted through the air.

6. What are the prime limitations on amplifier spacing and regenerator spacing? How do the two affect each other?

Answer: Amplifier spacing is limited by fiber attenuation; regenerator spacing is limited by noise and dispersion. Noise levels increase as amplifier spacing increases, so systems with long repeater spacing typically have shorter amplifier spacing.

ANSWERS TO QUIZ FOR Chapter 19

1. c.

2. b.

3. a.

4. d.

5. a.

6. e.

7. a.

8. c.

9. e.

10. c.

11. a.

12. b.

<u>Supplementary Materials</u>

Network maps are useful are useful in showing structure of the telecommunications network. Some are included in Chapters 23 and 24; the Bookmarks file on the CD-ROM includes some pointers

The books listed under "Further Reading" should help students who want more information on electronic communications systems and telecommunications. Gary M. Miller, *Modern Electronic Communications* (Prentice Hall, 1999) is at an introductory level suitable for most students.

Chapter 20:

———

Fiber System Standards

This chapter covers communication system standards relevant to fiber optics. It explains the concept of layered standards as well as specific transmission standards. The emphasis is on high-speed standards for backbone telecommunications, to prepare students for learning about system design and operation. Appendices C and D tabulate standard data rates and wavelengths used in fiber-optic systems.

Overview

The chapter opens with a brief introduction explaining the need for communication standards and how they have evolved. I draw an analogy between standard formats and agreements to speak a common language at a conference, and mention the fact that a 50-year old phone will still work with modern household telephone wires. (You can't count on it working in a newly wired office building or campus, however.)

A brief section explains families of standards, such as the hierarchy of data rates in time-division multiplexing. The goal is to teach students that different types of signals have different requirements, leading to different standards.

The next section covers layering of standards, which is an important and potentially confusing concept. Figure 20.1 gives one view—standard interfaces between layers so different layers "see" specific functions. Figure 20.2 shows the standard layers used for communication services. You may want to tailor your analogies to the background of your students. My goal is to teach them the basic concept, without digging into potentially confusing details.

A section on transmission formats covers how information is coded and packaged. It compares and contrasts time-division multiplexing and packet switching to help students understand the differences between standards aimed at the two applications.

Interchange Standards are next, with brief descriptions of Asynchronous Transfer Mode (ATM) and Internet Protocol (IP) standards. You should stress that although ATM cells may look like packets, they are designed for circuit switching so the format guarantees transmission capacity.

The section on fiber transmission standards covers physical layer standards—that is, how the data is packaged on the physical medium carrying signals. My emphasis is on fiber standards, but many of these standards can be used for transmission over copper over short distances or at low data rates. This section does not go into great detail, but does explain the basic functions of each standard. Later chapters give more detail on some individual standards.

The final section covers current issues and standards under development, such as those for optical networking and for digital video (covered also in Chapter 27). This section also describes how market pressures can affect standards.

Questions to Think About for Chapter 20

1. Follow the voice signals in Figure 20.2 through the layers in the diagram. What function does Internet Protocol serve?

Answer: Data packaging, it's an interchange format.

71

2. Many makers of telecommunications equipment are proposing to transmit IP signals directly on fiber, without going through ATM or SONET coding. What advantages might this have?

Answer: By eliminating layers, it would reduce equipment costs.

3. What difference between ATM and IPv4 formats is most important for voice transmission?

Answer: ATM has provisions for priority codes, so voice signals can be given high priority, simulating a virtual circuit. IPv4 lacks priority codes. (Note that IPv6 does have priority codes, but is not yet implemented widely.)

4. A major advantage of packet switching is that it can combine signals which arrive at uneven rates to use transmission capacity efficiently. Suppose you have 4 packet-switched input signals, which can arrive at peak rates of 1 Gbit/s. However, on average the packets account for only about 20% of the peak capacity. If all goes well, can you squeeze those four input channels through a 1 Gbit/s output?

Answer: Yes. The average data rate on each of the four inputs is 200 Mbit/s, so their average total is 800 Mbit/s. That should fit through a 1 Gbit/s output, although normal load fluctuations could cause delays at some times.

5. You can pack 24 voice channels on one T1 carrier, 4 T1 carriers into a T2 channel, and 7 T2 carriers into a T3 signal. How many voice channels can a T3 signal carry?

Answer: Multiply the numbers : 24 × 4 × 7 = 672

6. How many voice channels can an OC-192 signal carry, assuming an OC-1 carrier transmits the equivalent of one T3 carrier?

Answer: Multiply the number on a T3 (672) by the multiple of the OC-1 carrier (192) to get 129,024.

<u>Answers to Quiz for Chapter 20</u>

1. e.

2. a.

3. c.

4. b.

5. b.

6. d.

7. a.

8. e

9. c.

10. a.

11. d.

12. c

Solved Quiz Problem

Question 7:

ATM cells contain 48 data bytes and a 5-byte header, for a total of 53 bytes. SONET Frames are 810 bytes long and include a 27-byte header. That leaves 783 bytes for data. The SONET data includes both headers and data payloads from ATM cells, 53 bytes per cell. Dividing 783 by 53 gives 14.77 ATM cells per frame, but only 14 will fit. Note this is oversimplified because it ignores reserved fields within the SONET frame.

Supplementary Material

Appendix C lists standard TDM data rates in the Digital Telephone Hierarchies for North America and Europe, and in SONET and SDH.

Appendix D lists the ITU standard wavelength grid for WDM, with 100-GHz spacing.

Major standards groups have web sites that are useful resources. Although some are listed under Further Reading, you may want to search the web for others. Some standards groups functionally disband after completing their work, often—but not always—leaving web sites behind. Some organizations such as the International Telecommunications Union do not post details of their standards because their income relies on sales of documents on standards.

Chapter 21:

Single-Channel System Design

Chapters 21 and 22 introduce the fundamental concepts of system design at a level appropriate for an introductory course. The goal is to show students how systems are designed, without going into great detail. To achieve that goal, I show calculations for a few simple cases.

I split the design material into two chapters to keep the topic manageable. This chapter covers design of a single-channel system, which carries only a single wavelength. It concentrates primarily on power and dispersion budgets, with sample calculations for a few different cases. Chapter 22 covers more complex optical networks and wavelength-division multiplexing.

Overview

The chapter opens with a list of important variables in system design, and a brief explanation of their importance.

The first long section covers power budgets from a system viewpoint. The first subsection covers how light is collected from transmitters, considering both single- and multimode fibers. Other subsections cover fiber attenuation, receiver sensitivity, other losses, system margin, and the role of optical amplifiers. Most of this material was taught in earlier component chapters; this chapter fits these concepts into the context of systems.

The next section works through the loss budgets for three sample systems, chosen to illustrate different cases. One is a short point-to-point link within a building, a second is a 300-km link between two rural telephone facilities, and the third is a local-area network with a central star coupler. They are illustrative rather than realistic. In two cases, I show designs that don't work and how to fix them.

The next section covers budgeting for transmission capacity, expanding on descriptions of dispersion and time response in earlier chapters. My approach is to calculate the time response for transmitter, fiber and receiver, and use those numbers to obtain a total time response. That time-response figure, Δt, is then used to estimate data rate or bandwidth. I run through calculations of fiber dispersion for three examples: a multimode link, a 300-km single-mode system, and a 1000-km single-mode system. Then I cover the system implications of dispersion compensation and transmitter and receiver characteristics.

The final section covers cost/performance trade-offs. These are somewhat qualitative, and intended both to stimulate and to introduce important concepts such as the use of dark fibers to provide future room for expansion.

Questions to Think About for Chapter 21

1. WHY IS THE RELATIVE light collection efficiency of fibers in decibels proportion to *20* times the log of the ratio of their diameters rather than 10 times?

Answer: Because light collection is proportional to the area of the core, and area is proportional to the square of the diameter.

2. Suppose the amplifiers in a transatlantic cable are limited to 12 dB of gain to limit noise. How far apart can they be spaced if the fiber attenuation averages 0.24 dB? You can neglect splices, and the system contains no connectors.

Answer: 50 km.

3. You are installing fiber-optic data link between a laboratory and a remote data-collection center 5 km away. You want to use a VCSEL transmitter with -5 dBm output and a fiber with loss of 2.5 dB/km at 850 nm. The system includes patch panels on both ends, with two connector pairs at each patch panel, and additional connectors on the terminal equipment. If connector loss is 0.5 dB, and you want a 10 dB margin, how sensitive must your receiver be?

Answer:

Loss Budget: Transmitter -5.0 dBm

Fiber 5 km x 2.5 dB/km 12.5 dB

Connectors: 6x0.5 3 dB

<u>*Margin 10 dB*</u>

Receiver sensitivity -30.5 dBm

4. You have to design a system with 1 Gbit/s data rate using return-to-zero (RZ) digital coding. What is the 3-dB analog bandwidth of this system? What NRZ data rate could this system transmit.

Answer: First calculate the pulse spreading allowable in a 1 Gbit/s RZ system; the answer is 0.35 ns (0.35 divided by 1 Gbit/s). Then use that number to estimate an analog bandwidth of 1 GHz (350 MHz divided by 0.35 ns). The NRZ data rate is twice the RZ, or 2 Gbit/s.

5. You want to transmit 1 Gbit/s through 100 km of single-mode fiber, using a transmitter and receiver that each have response times of 0.4 ns. The transmitter has line width of 0.1 nm. Neglecting polarization-mode dispersion, what is the maximum chromatic dispersion allowable in the fiber.

Answer: First calculate the total response time allowable, 0.7 ns. Then use that to calculate the allowable response time of the fiber. This starts by plugging the known numbers into the equation for response time

$$0.7 \ ns = \sqrt{0.4^2 + 0.4^2 + \Delta t_{fiber}^2}$$

and going back to calculate the unknown fiber response time Δt_fiber which comes to 0.41 ns. Then work backward to calculate the maximum allowable chromatic dispersion in the fiber

0.41 ns = 100 km x 0.1 nm x D ps/km-nm

The fiber's characteristic dispersion D must be 41 ps/km-nm or less. That's an easy figure to achieve.

6. Can you get away with using a 1550 nm VCSEL in the system of problem 5 if the VCSEL has linewidth of 0.5 nm, output of -5 dBm, and the fiber loss is 0.25 dBm? (This assumes you can find such a VCSEL.) Check both the pulse spreading and the power level, assuming receiver sensitivity of -30 dBm.

Answer: The dispersion would have to be five times lower, 8.2 ps/km-nm, which is possible with non-zero dispersion shifted fiber.

The power budget is

Transmitter output -5 dBm

Fiber loss 0.25 dB/km x 100 km -25 dB

Receiver sensitivity -30 dBm

Margin 0 dB

The low power of the VCSEL leaves no power margin with a -30 dBm receiver at 100 km. This isn't a feasible system. You need either shorter distance, a more powerful transmitter, a more sensitive receiver, or an optical amplifier.

Answers to Quiz for Chapter 21:

1. c

2. c

3. d

4. c

5. b

6. c

7. c

8. e

9. d

10. d

11. d.

12. c.

Solved Quiz Problems

Problem 5.

The two components of the loss budget for each span between optical amplifiers are fiber attenuation and splice loss. With one splice every 16 km, there are 4 splices along an 80-km span. The loss budget is:

$80 \text{ km} \times 0.3 \text{ dB/km} = 24.0 \text{ dB}$

$4 \times 0.1 \text{ dB/splice} = 0.4 \text{ dB}$

Total span loss 24.4 dB

Problem 6.

To calculate the loss budget, consider the loss to the furthest home, 4 km from the transmitter.

4 km × 0.4 dB/km 1.6 dB fiber loss

6 connectors I 0.5 dB 3.0 dB connector loss

1×200 coupler 23.0 dB coupler loss

System margin 5.0 dB

Total attenuation 32.6 dB

Receiver sensitivity -30.0 dBm

Transmitter power 2.6 dBm

Problem 8.

Calculate response time using the sum of the squares law for transmitter, receiver, and fiber response. 100 m of fiber with 20 ns/km dispersion has total dispersion of 2 ns.

$$t = \sqrt{t_{tr}^2 + t_{rec}^2 + t_{fiber}^2} = \sqrt{2^2 + 1^2 + 2^2} = 3\ ns$$

Problem 9.

First calculate the modal and chromatic dispersions separately, then combine them by the sum of the squares rule.

Modal dispersion: 10 km × 2.5 ns/km = 25 ns

Chromatic dispersion: 10 km × 100 ps/nm-km X 50 nm = 50 ns

$$t = \sqrt{\left(\Delta t_{modal}\right)^2 + \left(\Delta t_{chromatic}\right)^2} = \sqrt{25^2 + 50^2} = 55.9\ ns$$

PROBLEM 10.

The problem is similar to 9, but the numbers are different, and different dispersions are involved.

Chromatic dispersion: 10 km × 17 ps/nm-km × 1 nm = 170 ps

$$PMD = \left(0.5\ ps/\sqrt{km}\right) \times \sqrt{10\ km} = 1.58\ ps$$

$$t = \sqrt{\left(\Delta t_{chromatic}\right)^2 + \left(\Delta t_{PMD}\right)^2} = \sqrt{170^2 + 1.58^2} = 170.007\ ps$$

Note that polarization mode dispersion is not always this insignificant. With a narrow-line (0.001-nm) source, it would be the dominant component of dispersion for this system. Note also that PMD is averaged out in this example, but in reality fluctuates continually in time as conditions vary.

<u>Supplementary Materials</u>

THE NATURE OF THIS chapter makes it a natural one for you to supplement with your own problems for students to calculate.

To help students appreciate the difficulties in trans-oceanic cable design, you can have them calculate dispersion requirements for a 5000-km cable at various speeds.

If you can get examples of system design from published articles, vendors, or other sources, you can work out real-world designs in front of the class.

Try a fiber to the home design.

Chapter 22

Optical Networking System Design

Wavelength-division multiplexing and optical network increase the complexity of a fiber-optic system as well as enhancing its transmission capacity. This chapter introduces major concepts involved in multi-wavelength systems, but does not go into detailed design examples. Design techniques are still evolving, and at this stage of development it's most important for students to understand concepts. This chapter has been rewritten for the post-bubble world.

Overview

The chapter opens with a brief summary of concepts that fall under the heading of optical networking, including transmission capacity, granularity, and switching.

The first section covers optical channel density and capacity-related issues. It compares the benefits of WDM with high-speed time-division multiplexing, and the ways of packing optical channels. It also points out that comparatively few optical channels in most systems have actually been populated with signals and are carrying live traffic. The capabilities and designs of coarse and dense WDM are compared, with the CWDM channel grid plotted in Figure 22.4.

The next section covers the wavelength ranges over which WDM systems operate. It focuses on two primary issues. One is the gain bandwidths of available optical amplifiers, which are required for long-distance transmission. The other is the transmission bandwidths of optical fibers, which can be limited by both fiber attenuation and by fiber dispersion.

The following section outlines important factors in WDM design, not counting amplifiers. One is fiber attenuation, which varies with wavelength. Dispersion slope and compensation are other issues, both of which also vary across transmission bands. Figure 22.8 shows how cumulative chromatic dispersion varies along the length of a dispersion-managed system. Nonlinear effects also are covered, particularly four-wave mixing, which is related to dispersion.

The impacts of optical amplification on WDM system design are covered next, including amplifier power levels, gain flatness, and channel equalization.

The important role of switching signals, both between fibers and between wavelengths, is covered in the next section. Optical networks can be classed as transparent or opaque, and all-optical or hybrid opto-electronic, but in practice most networks are hybrids, containing some optical segments and others that include electronic segments. The section looks to the long term and describes the need for wavelength conversion, but in practice requirements are limited today.

The final section includes two brief design examples. One considers amplifier gain and system power requirements. The other looks at the impact of adding 10 optical channels to an existing system. These simple choices will give the student a feel for optical networking, without delving into the serious complexities of trying to achieve high-speed optical networking over long distances, which are beyond the scope of an introductory text.

<u>Questions to Think About for Chapter 22</u>

1. YOUR DESIGN ALLOWS you to pack 2.5 Gbit/s optical channels only 50 GHz apart, but 10 Gbit/s signals require 100 GHz spacing to give adequate performance margins. Which allows you to transmit more data if you populate every available slot in the erbium-amplifier C band? By how much?

Answer: More widely separated 10-Gbit/s signals can transmit at twice the total data rate because they give 10 Gbit/s per 100 GHz, compared to only 5 Gbit/s per 100 GHz for the 2.5 Gbit/s signals.

2. You have wavelengths of 1460 to 1625 nm available in an unamplified metro WDM system. How many CWDM optical channels can you transmit with ITU standard 20-nm spacing?

Answer: You can transmit the eight ITU channels at 1470, 1490, 1510, 1530, 1550, 1570, 1590, and 1610 nm.

3. You have two types of fiber available for dispersion compensation, step-index single-mode fiber with dispersion of +17 ps/nm-km at 1550 nm, and a reduced-slope nonzero dispersion-shifted fiber with -2 ps/nm-km at the same wavelength. How much of each do you need for a 95-km span with zero cumulative dispersion at 1550 nm? Which should you use closer to the transmitter?

Answer: You need 2 km of the step-index fiber for every 17 km of the reduced-slope fiber, a total of 10 km of step-index fiber with total dispersion of +170 ps/nm, and 85 km of reduced-slope fiber with total dispersion of -170 ps/nm. The step-index should go first because it has a larger effective area, which would reduce nonlinear effects at the higher powers near the transmitter.

4. An erbium-fiber amplifier has a peak output of 20 dBm. You use it to transmit 40 optical channels. What is the output power on each? What happens to the output power per channel if you double the number of channels?

Answer: You have to assume the power is divided equally among the channels. Divide 100 mW by 40 channels, yielding 2.5 mW per channel. Doubling the number of channels reduces power per channel by half to 1.25 mW.

5. You need to send 20 optical channels through a series of 50 erbium-fiber amplifiers, with the output signals differing by no more than 6 dB at the end of the amplifier chain. How precisely do you need to equalize gain if all that difference is due to unequal gain across the range of optical channels?

Answer: Divide 6 dB by 50, giving 0.12 dB.

6. The input to an all-optical switch is a cable containing 48 fibers. Each fiber can transmit up to 40 optical channels. How many optical channels can the switch direct if it has 128 input and 128 output ports?

Answer: This is a trick question. The switch has enough ports to handle more than 48 fibers, so it could handle all the input fibers by switching all optical channels in each fiber, a total of 1920 optical channels. If all the optical channels were demultiplexed before switching, it could handle only 128 optical channels—but the question didn't say the channels were demultiplexed. It is true to say the switch could handle 1920 optical channels, but only by directing multiple channels along the same routes. If you wanted to switch channels individually, the right answer is 128. If your job is to sell switches, you are likely to say 1920. This example shows how different points of view can lead to different specifications.

<u>Answers to Quiz for Chapter 22</u>

1. c.

2. d.

3. a.

4. d.

5. b.

6. c.

7. a.

8. d.

9. c.

10. a.

11. c.

12. b.

<u>Supplementary Material</u>

Countless articles were published over the past several years on designs for optical networks that would in theory provide tremendous potential capacity. Unfortunately many provided little information on how the networks would operate, and most were written to sell ideas rather than critically evaluate concepts. Be careful, particularly of trade magazines.

The best overviews of system concepts and trends often come in tutorial papers at two major fiber optics meetings—the Optical Fiber Communications Conference and the European Conference on Optical Communications. Some appear in conference digests as series of slides which can be very helpful.

Two books that deal specifically with optical networking concepts are:

Vivek Alwayn, *Optical Network Design and Implementation* (Cisco Press, Indianapolis, 2004)

Rajiv Ramaswami and Kumar N. Sivarajan, *Optical Networks: A Practical Perspective, 2nd ed* (Morgan Kaufmann, San Francisco, 2002)

Chapter 23:

Global Telecommunications Applications

This chapter serves two purposes: to introduce the global telecommunications network, and to describe the network of submarine cables that links all continents but Antarctica. Chapters 24-27 cover other parts of the global telecommunications network. These chapters are application-oriented, and concentrate on systems.

Overview

The first step in understanding the global telecommunications network is to define telecommunications. The first section of this chapter does this, then looks at the various types of telecommunications and the notion of "convergence" which is bringing together systems originally designed for distinct purposes. Subsections briefly describe the three most pervasive telecommunications networks, the telephone system, cable television, and the Internet, noting their connections and differences.

The next section explains how these networks collectively form a global telecommunications network, with extensive interconnections. It also explains the major transmission hierarchies that were built upon the telephone framework

•The time-division-multiplexed plesiochronous digital telephone hierarchy that steps from single phone lines to T3 rates.

•The European/ITU equivalent

•SONET and SDH formats from 52 Mbit/s to 10 Gbit/s.

The following section covers Internet transmission systems and the rapid growth of Internet traffic. It describes how the network was massively overbuilt during the telecommunications bubble, and a "thing to think about" tells about the glut of fiber transmission capacity in many long-haul networks.

The next section describes submarine cables, the backbones of international communications. I give a bit of historical background on submarine telegraph cables, and early submarine telephone cables, to show the tremendous advantages of fiber over other cables and over satellite links for telephone transmission. The global map in Figure 23.6 illustrates the extent of connections. Next I describe submarine cable technologies, and the various types of submarine cables—from short unrepeatered systems to trans-oceanic cables thousands of kilometers long. Maps illustrate some samples. Table 23.1 compares the technologies used in six submarine fiber cable systems, showing the steady advances in capacity. Note that cable capacity has reached the point where all channels are not initially populated in the most recent cables, Pan-American Crossing and Apollo.

The following section covers the long-distance terrestrial networks that are the continental counterparts of submarine cables. Figures 23.11 and 23.12 compare Qwest's fiber-optic backbone in North America with its Internet Protocol network. Both run through the same fibers, but the IP network is on a different layer, and shows only the routing notes where signals are transferred to the IP layer. This section explains the functional differences between terrestrial cables, which have to interconnect widely distributed population centers on land, and submarine cables, which serve fewer nodes on islands and coasts. I also mention the importance of add/drop multiplexing and wavelength conversion.

A final brief section explains the differences among the public switched telecommunication network, the Internet, and private leased lines.

Questions to Think About for Chapter 23

1. The data rate of a SONET OC-12 carrier is 622.08 Mbit/s. If that corresponds to 8064 voice channels, and each voice channel is 64,000 bits per second, how much of that signal is overhead?

Answer. Multiply 8064 by 64,000 and you get a data rate of 516.096 Mbit/s. Thus about 106 Mbit/s, or 17% of the OC-12 signal, is overhead. Alternatively, the overhead amounts to about 20.5% of the data rate of the actual voice channels.

2. SONET frames are a fixed length. Internet packets are a variable length, with the header indicating the packet size. How does this difference relate to the difference between circuit and packet switching?

Answer: The SONET protocol is equivalent to circuit switching because it allocates a fixed capacity (frames per second, which corresponds to bits per second) to signals. Packet switching does not guarantee a fixed capacity, so packets do not have to be equal length.

3. TAT-8 transmits 560 Mbit/s and began operation at the end of 1988. Atlantic Crossing 1 began operation in 1998 transmitting 40 Gbit/s on a parallel route. How much did data rate increase over that decade?

Answer: Dividing gives an increase of 71.4 times.

4. Remember that "Moore's Law" says that the capacity of integrated circuits doubles every 18 months. Judging from the increase in data rates on transatlantic cables in Question 3, what was the doubling time of fiber-optic capacity from 1988 to 1998?

Answer: The factor of 71.4 increase equals 2^x where x is the total number of times capacity has doubled in 10 years. Calculate x by taking the log of both sides (remembering that log (a^x)=x log(a))

$$\log(71.4) = \log(2^x) = x\log(2)$$

or $$x = \frac{\log(71.4)}{\log(2)} = 6.16$$

Divide that number of doublings into the time span of 10 years, and you get a doubling time of 1.62 year, surprisingly close to Moore's Law.

———————

5. USING THE RESULTS of Questions 3 and 4, if fiber-optic capacity continued to expand at the same capacity, what would the capacity of a transatlantic fiber-optic cable be in 2005? How does that compare with the Apollo Cable as it began operation in 2003? How does it compare with the potential capacity if all possible optical channels were activated?

Answer: That's 7 years from 1998, so calculate how many doubling periods 7 years represents. The answer is 4.23, Raise 2 to that power and you predict an increase by a factor of 20, to 800 Gbit/s. The Apollo Cable had 16 channels at 10 Gbit/s on each of four fiber pairs, a total of 640 Gbit/s running, when it began operation in February 2003, which is a little better than the historic trend. Its total potential capacity is 80 channels at 10 Gbit/s on each of four fiber pairs, or 3.2 Tbit/s. That capacity is theoretically available in 2005, but with the fiber glut the extra channels will not be populated with transmitters and receivers because the capacity is not needed.

6. In 1983, the peak data transmission rate in a terrestrial fiber-optic cable was 400 Mbit/s on a single fiber. In 2000, you could buy a system capable of transmitting 160 optical channels at 10 Gbit/s through a single fiber. How much an increase does that represent, and what's the doubling time?

Answer: The increase is a factor of 4000. That corresponds to 11.97 doublings, or 2^{12} over a period of 17 years, or a doubling every 1.42 years.

7. In 2003, the laboratory record for highest data rate through a single optical fiber was 10 Tbit/s. If the state of the art in 1983 was 400 Mbit/s, calculate how much of an increase that represents and what the doubling time is?

Answer: The increase is a factor of 25,000 over 20 years. Using the same set of calculations in Question 4, we find the number of doublings x is

$$x = \frac{\log(25,000)}{\log(2)} = 14.61$$

Since that is over 20 years, the time per doubling is 1.37 years, better than Moore's Law. It's also much better than anybody needs, because nobody is installing systems at even 1 Tbit/s.

<u>Answers to Quiz for Chapter 23</u>

1. e

2. d

3. a.

4. d

5. a.

6. e.

7. c.

8. b.

9. e.

10. c.

11. c.

12. a.

13. b.

14. b.

15. b.

<u>Supplementary Material</u>

MANY OTHER BOOKS DESCRIBE the global telecommunications network in more detail; you can refer interested students to them. If anyone is interested in the story of the first submarine telegraph cables, Arthur C. Clarke tells an enchanting tale in the first part of *How the World Was One* (Bantam, New York, 1992).

It's sometimes possible to get wall charts showing submarine cable routes from the companies that make or operate submarine cable, such as Alcatel and Tyco Submarine Systems. Some market-research companies also offer their own maps.

You can find extensive resources on cable routes and links to many on-line maps at http://www.cybergeography.org. As with other web sites, links can get outdated, so you should check before using it in class, but it's a great resource.

Overbuilding of fiber-optic networks was particularly intense on major submarine cable routes such as across the north Atlantic. The market-research firm TeleGeography has published its own tabulations, which are interesting and might be useful for discussions.

A good subject for class discussion is how much capacity is really needed. The record transmission capacity on a single fiber is around 10 Tbit/s. It's hard to appreciate how much capacity this represents until you break this down into 64,000 bit/s voice channels. If you do the arithmetic, you'll find at that rate you could put half the adult population of the US on each end of the fiber and—if you had enough phones to go around—everyone on one end could talk to someone on the other end through that fiber. (For a two-way conversation, you'd need a pair of fibers, and the numbers aren't quite enough to handle the whole population of the U.S., but it's still a staggering thought.)

Chapter 24:

Regional and Metro Telecommunications

This chapter covers both regional and metro networks, defined loosely as the links between the national long-distance network and the local distribution system. These networks evolved from regional telephone systems, and telephone service remains a major application, but thanks to convergence metro and regional networks also carry other signals, including Internet traffic. I include some explanation of industry structure so students can relate the textbook examples to what they see in the real world. This chapter is relatively brief, and you may want to cover it along with national long-distance networks.

Overview

I open with the general concept of a regional network, which I consider to include metropolitan and suburban networks as well as rural ones. Three examples show the concept. The first is an imaginary rural area, with links from small and large towns to a small city that serves as a regional hub. The second is an actual regional data network operating in Oregon. The third is an imaginary network serving a large city and suburbs.

The next section describes established regional telecommunication networks—based on the traditional dominant telephone carriers in each region. I hesitate to call them "Baby Bells" after the past few years of corporate mergers, where I've seen the dominant carrier in the Boston area change its identity from Nynex to Bell Atlantic to Verizon (which includes remains of the former GTE). This section gives some of this background and sketches the structure of existing networks.

The following section introduces the new vision of "metro" networks that serve a somewhat different purpose. Typically these are loops reaching from city to suburbs, installed by companies other than the traditional phone companies, and leasing fiber capacity to all comers—including traditional phone companies which need more capacity between their facilities.

The rest of the chapter covers design features of these moderate-distance networks. I show how add/drops can provide services to individual nodes, discuss transmission speed requirements, and show how WDM is used. I also describe the distance scales and the choices that implies for operating wavelength, fiber choice, and optical amplifiers.

Questions to Think About for Chapter 24

1. Telephone calls that go outside of your community pass between your local telephone switching office and other switching offices on the regional telephone network. What makes some of these long-distance calls?

Answer: Regulations and rate structures are the short answer. From a regulatory viewpoint, long-distance calls are made between "local area and transport areas" (LATAs) defined at the time AT&T broke up the first time in 1984 into seven regional phone companies and one long-distance company. These zones were based on area codes that existed at the time, but most of those area codes have been split into two or more area codes to provide more phone numbers. For many years, the dominant carrier in each region (the former Bell Operating Company) could not provide long distance service, and AT&T likewise could not provide local phone service. That's changing slowly. The rate structures that define what are toll calls (charged in addition to your basic monthly rate) are charges set by phone companies and local public utility commissions. Typically different flat rates cover calls to different areas. Note also that in some locations you may pay toll charges to call numbers in the same area code, while in others numbers with different area codes may be local calls.

2. A fire destroys the local switching office in one of the "larger towns" of Figure 24.1. How can the smaller towns connected to it still receive phone service?

Answer: Through other lines in the mesh of the regional network.

3. Your company needs more transmission capacity between two offices 20 km apart. You can lease wavelengths at $20,000 per wavelength per year on a metro network with a five year lease. You need to transmit three channels, two containing gigabit Ethernet, and one containing TDM phone signals at 622 Mbit/s. They could fit on a single 2.5 Gbit/s line, but you would need to buy a time-division multiplexer and demultiplexer for the job. What's the most you could pay if you have to amortize costs over 5 years.

Answer: You're trying to reduce your requirements from three optical channels to one. That would save you $200,000 over five years. You would have to find a multiplexer and demultiplexer that cost under $200,000 to make it worthwhile.

4. You're dealing with the same metro network as in Question 3. Their base rate is $1000 per kilometer per wavelength per year. What's the most you could pay for a time-division multiplexer and demultiplexer if you were sending signals only 5 km?

Answer: $50,000. Rates aren't always this simple, but this highlights how transmission distances can affect economics. It's worth noting that most current long-distance rates in the United States no longer depend on distance—all calls outside the local area are charged at the same rate per minute.

5. A metro network uses low-water single-mode fiber with zero dispersion at 1310 nm, and dispersion of 17 ps/nm-km at 1550 nm. If it transmits signals 100 km, what is the maximum data rate it can handle in the 1550-nm window with a transmitter having linewidth of 0.1 nm. Assume polarization mode dispersion is insignificant, and that the maximum transmission speed is

$$Data\ rate = \frac{0.7}{\Delta t_{max}}$$

Answer: Calculate the time spreading due to chromatic dispersion

$$\Delta t_{chromatic} = \left(17\ ps/nm - km\right) \times 100\ km \times 0.1\ nm = 170\ ps$$

Plug the resulting time response, 170 ps, into the data rate formula, and you get 4 Gbit/s. Thus the system could easily transmit a 2.5 Gbit/s signal even at a wavelength where fiber dispersion is high.

6. You're using non-zero dispersion-shifted fiber in a metro network. If the chromatic dispersion is 1 ps/nm-km and the source bandwidth is 0.1 nm, how far can you transmit signals at 40 Gbit/s? Assume you can neglect polarization-mode dispersion.

Answer: The total allowable dispersion is 17.5 ps, so the maximum distance to meet the data rate criteria is 175 km.

<u>Answers to Quiz for Chapter 24</u>

1. c.

2. a.

3. d.

4. c.

5. e.

6. f.

7. a.

8. b.

9. b.

10. d.

11. b

12. d.

<u>Supplementary Material</u>

A free web site http://www.cybergeography.org has great resources on cable routes and links to many on-line maps, including some regional ones.

Regional route maps are hard to get from telephone companies, but may be available from state or local utility officials.

The division between regional and long-distance service was made at the time of the first AT&T breakup in 1984, based on area codes that existed then. Your students probably have no idea what areas those area codes covered then. It's worth digging up an old map of area codes from that period; I found one from the 1970s at **http://www.lincmad.com/ map1970s.html**

Chapter 25:

———

Local Telephone or "Access" Networks

The local telephone network runs from the local switching office to the subscriber's phone, and is the part of the network most visible to customers. Traditionally fiber has made up only a small part of this network, with fiber lines running directly only to a few businesses. However fiber is now being deployed much more widely, and some large phone companies are building networks that bring fibers all the way to homes. Although I call this a "telephone" network, it is more generally called an "access" network, which also carries both high-speed digital lines for business customers and broadband services to homes.

<u>Overview</u>

The first section describes the structure of the local network, introducing important terms such as "subscriber loop," and "access." It introduces the concept of the network edge as an interface between the larger-scale network and local distribution services where calls are directed. Then it shows how telephone switching offices provide this function in conventional telephone networks. I also describe "access" customers on the edge of the network, and describe the operation of cellular telephone systems. This section closes with an illustration of how cables distribute signals through the subscriber loop.

The next section explains the services offered through the network, starting with leased lines for business customers, which often go over optical fibers. I explain the digitization of voice phone lines, and how standard copper wire pairs can carry Digital Subscriber Line (DSL) signals. Table 25.1 lists the types of digital subscriber line services available and the maximum distance spanned. This section also described other emerging services such as voice over Internet Protocol (VoIP) and technologies that in some way compete with fiber for delivering broadband services.

The next section gives an overview of emerging services and competing technologies. , which are using fiber to deliver services closer to homes. The family is sometimes called "fiber to the X" (FTTX), where the "X" means anything from "neighborhood" to "home" or "premises." Today's telephone networks typically use fiber to deliver services to neighborhood nodes. Fiber to the Curb (FTTC) delivers fiber to a box on the curb that serves homes on that block. Fiber to the Home began with grass-roots projects in small cities, towns and rural areas, often operated by municipal utilities, and with private networks installed by developers of large subdivisions. In 2004 Verizon announced it would install fiber past a million homes by the end of that year, with more planned for 2005. This time they're for real—I can see the fiber on the poles across the street from my home, although they haven't started signing up customers.

The final two sections cover the two leading approaches to delivering fiber to the home: passive optical networks and Gigabit Ethernet. Both have their advantages and you should ask your students to compare them.

———

<u>Questions to Think About for Chapter 25</u>

1. A SWITCHING OFFICE serves 5000 voice telephone lines. It is designed so that at peak usage 20% of the lines can be connected—a total of 1000 phone lines. Residents discover the Internet and 500 of them buy dial-up modems and

install new phone lines so they can leave their modems on all of the time. How much more switching capacity does the phone company have to install—both in number of lines and percent?

Answer: If all the modem lines are connected all the time, the phone company needs 500 more outgoing lines, an increase of 50% to a total of 1500 outgoing lines.

2. Suppose that instead of installing new phone lines, the residents of the town in Question 1 hooked their modems up to existing phone lines. If 500 people buy modems and half of them go on the Internet in the evening at a time of peak residential calling, how much does the phone company have to increase its switching capacity? How much must switching capacity be increased to accommodate half the households buying modems and half of the modem users going on the Internet in the evening? Assume the rate of voice calling does not change.

Answer: If half of 500 people with modems are on the Internet, they require 250 switched circuits. Of the remaining 4500 people in the town, 20% or 900 are assumed to be making voice calls, so the total capacity needed is 1150 circuits. That's a 15% increase although only 10% of the people have bought modems. If half the population has modems and half of the modem users are on the Internet, they require 1125 circuits just for computers. The remaining half include 20% on voice lines, or an extra 500 circuits—for a total of 1625 voice circuits.

3. Check to find how far your residential phone line is from your local phone company's switching office at *http://www.dslreports.com*. What DSL rates are available? If you're in a class, compare the rates and distance with those of other students.

Answers: will vary. Students will learn that cable runs are not "as the crow flies," because cables and conduits follow the courses of streets.

4. A fiber to the curb system is installed on a street where it serves 10 homes with copper drop cables less than 1000 feet long which carry one phone line plus VDSL service at the maximum possible speed. If all voice and data services to that curbside interface are digitized and transmitted on a fiber, what is the minimum data rate on the fiber to serve all homes.

Answer: VDSL requires 52 Mbit/s to each of 10 homes, a total of 520 Mbit/s. Voice phone requires 56 kbit/s for each of the 10 homes, a total of 560 kbit/s. The total data rate is 521 Mbit/s, so you need a 622 Mbit/s OC-12 carrier.

5. A passive optical network serves 32 subscribers with 622 Mbit/s downstream and 155 Mbit/s upstream. Two customers buy premium service which guarantees 50 Mbit/s downstream and 10 Mbit/s upstream. What speed can the other customers get if the remaining capacity is divided equally among them?

Answer: The remaining capacity downstream is 622 - 100 Mbit/s, or 522 Mbit/s, divided among 30 customers, who each receive 17.4 Mbit/s. The remaining capacity upstream is 155-20 Mbit/s or 135 Mbit/s, which divided among 30 customers comes to 4.5 Mbit/s upstream for each. Note that some schemes would allow flexible allocation, so customers would have a guaranteed minimum capacity, with the unused excess shared among customers who needed it at the time.

6. The transmitter for a passive optical network generates a 1-mW signal that is divided equally among 32 users. If the cable loss is 10 dB and the couplers have no excess loss, what is the signal that reaches each user?

Answer: After 10 dB attenuation, a 1-mW signal becomes 100 μW. Dividing that among 32 users gives each 3.2 mW.

7. A Gigabit Ethernet signal is split among 32 subscribers. Neglecting losses arising from congestion, what is the maximum data rate if all are receiving signals at equal capacity?

Answer: Divide 1 Gbit/s by 32 to give 32 Mbit/s. This is not realistic because congestion would increase losses.

<u>Answers to Quiz For Chapter 25</u>

1. d.

2. a.

3. b.

4. b.

5. c.

6. b.

7. e.

8. b.

9. c.

10. c.

11. a.

12. b.

13. e.

14. c.

15. e.

<u>Supplementary Material</u>

Your local phone company, or public utility commission, may be able to tell you something about the phone service to your area, but don't expect the phone company to supply route maps.

Your campus IT department should be able to give you some information on the campus network.

For more on Passive Optical Networks, see **http://www.fsanweb.org/**

Chapter 26:

Computers and Local-Area Networks

Previous chapters have considered data as part of the traffic on the global telecommunications network. This chapter looks specifically at data transmission, packet switching, the Internet, and local-area networks. It emphasizes data transmission separate from the overall telecommunications network, but also mentions interfaces with the overall network. Although I give some general background on local area networks, this chapter is about fiber networks, not LANs in general.

Overview

The opening section compares telephone and data transmission. An important contrast is in the flow of traffic; computer data transmission is bursty, while voice is more even. This leads to the central dichotomy between packet switching of data and circuit switching of voice signals. The section also illustrates the layering of computer network protocols involved in activities such as electronic mail.

The next section describes Internet structure, starting with the global backbone network and points of presence, then moving to Internet Service Providers and more local systems. Remember to emphasize that Internet maps such as Figure 26.4 show logical rather than physical links between points of presence. The curved lines show that connections exist between nodes at points such as Phoenix and Dallas, but do not show the path of the cables, which are on the physical layer.

The rest of the section moves to successively smaller scales. It describes metro and wide-area networks, then local-area networks, noting that some terminology can be ambiguous. Local-area networks may not be familiar to all students; if they use your campus LAN, it would make a good example. The section also describes point-to-point links between computers and individual peripherals, and optical interconnection within computers and other equipment.

The next section compares data transmission technologies. One consideration is the network topology, such as star, switched, or broadcast. Another is the transmission technology, including copper, fiber, or wireless links. Fiber's advantages over copper include much greater immunity to electromagnetic interference and ground loops. This section also describes wireless data networks and free-space optical networks.

A brief section reviews the types of fibers, transmitters, and receivers used in fiber data links.

The final section covers applications of fibers in standard data networks. I picked standard 10-Mbit/s Ethernet as the starting point to discuss Ethernet, although fiber is rarely used at that speed. Fibers are used at higher speeds, and are integral parts of the standards for Gigabit Ethernet and 10-Gigabit Ethernet. Tables list fiber transmission formats for Gigabit Ethernet and 10-Gigabit Ethernet.

Questions to Think About for Chapter 26

1. IF THE INTERNET is the computer equivalent of the long-distance backbone system for telephony, what is a good analogy for the regional telecommunications network?

Answer: The best answer would be the metro network or MAN. A WAN is an acceptable answer because the term is so vague.

2. A cable modem shares a loop of cable with other subscribers in your neighborhood, in effect making all of you share a local-area network run by the cable company. Suppose that the modem is able to deliver 10 Mbit/s to your computer's Ethernet port if nobody else is online. Your next-door neighbor signs up after you say how great it is, then gets eight more neighbors to sign up. If the capacity available to you depends only on the number of users, how much of the original capacity is left?

Answer: With 10 users, each would nominally get 10%. In reality, this would vary considerably, depending on the load and the usage level of each neighbor. At peak loads, collisions reduce total efficiency of Ethernet, so total speeds would drop much lower. If nobody else was on line or if your neighbors usage was low, you wouldn't notice a difference.

3. A fiberless optical network can transmit signals between any two office buildings 95% of the time on foggy days. If you have a single laser link to your building, how much time would you expect to be down during a 24-hour interval of foggy weather? Suppose you could add two more laser links that are completely independent of the first and of each other, and would connect you to the same network. How much would your service improve?

Answer: 95% reliability means the system would be down 5% of 24 hours, which is 72 minutes, a long enough outage to be a serious problem. If you have two more independent links. you only need one of the three to work. the odds of all three failing are 0.05 × 0.05 × 0.05 or 0.0125%, or 0.18 minute. In reality, the probability of failure is not likely to be completely independent because dense patches of fog are likely to obstruct more than one path to your building.

4. Your company has just signed a 10-year lease on a building being renovated. The local-area network in your old building was a standard 10 Mbit/s Ethernet, but traffic was doubling every year and the network is at capacity. You want to stay with Ethernet standards. How long can you go before you run out of capacity on Fast Ethernet? How long before even Gigabit Ethernet won't do?

Answer: The number of years N required to increase a factor of 10 is given by $2^n=10$. Taking the log of both sides gives n=(log 10)/(log 2)=3.3 years. At the same rate, it will take only another 3.3 years to increase another factor of 10, so in 6.6 years you run out of capacity on Gigabit Ethernet. That means you should prepare for 10-Gigabit Ethernet.

5. You need to transmit Gigabit Ethernet between two buildings separated by 400 m. What are your options? What are your options if you expect to have to upgrade to 10-Gigabit Ethernet?

Answer: Graded-index 50/125 or 62.5/125 fiber with high bandwidth can transmit Gigabit Ethernet over that distance at 1310 nm, and high-bandwidth 50/125 fiber can transmit that distance at 850 nm. However, only single-mode fiber can transmit 10 Gbit/s Ethernet that distance.

6. Why would you *not* want to use DWDM to increase transmission capacity of a local-area network?

Answer: The WDM optics and multiple transmitters and receivers are quite expensive. For the short distances involved, it would be much less expensive to lay additional fiber.

7. Why is the Internet sometimes called a wide-area network?

Answer: Because it interconnects local-area networks and other smaller-scale networks.

<u>Answers to Quiz for Chapter 26</u>

1. c.

2. b.

3. d

4. a

5. b

6. c

7. b

8. b

9. c

10. d

11. e

12. b

13. d

14. a

<u>Supplementary Materials</u>

YOUR CAMPUS NETWORK is a good starting point for discussing data network technology.

Many standards organizations have their own web sites, which give technical information as well as plugging their products. Unfortunately, some of them have disappeared or were not on-line when I compiled this guide.

The Fibre Channel Industry Association web site is **http://www.fibrechannel.org**

Chapter 27:

Video Transmission

Video transmission is less integrated with the global telecommunications network than voice or data. Cable television networks account for most uses of fiber in video, although there are other applications. The big change coming in video is the advent of high-definition and digital television, but it's taken much longer than expected to get off the ground, and will undoubtedly cause significant disruption.

<u>Overview</u>

I open with a long section introducing the basics of video transmission, starting with scanning formats, and including both analog and digital transmission standards. The emphasis is on American standards, but I also mention some other standards. I spell out key differences, but do not go into great detail because this is not a video course. I also explain how display standards affect computer monitors. I also explain the distinction between "digital television" as manifested by DVD players and satellite systems, and high-definition digital television, or HDTV, which offers higher resolution. As Table 27.2 shows, digital is not only "high-definition" television. A box describes the muddled state of HDTV as this book went to press.

The brief section that follows outlines the major ways of distributing television signals: terrestrial broadcast, satellite broadcast, terrestrial microwaves, and cable television.

The next section describes the architecture of cable television networks, the major video application of fiber optics. After describing the architecture of a cable system, I describe the widely used hybrid fiber/coax design. Like local telephone networks, cable-television systems generally use fiber to distribute signals to nodes, from which signals are distributed over copper. This section also explains the operation of cable modems and telephone over cable. The goal is to help the student understand the operation of the cable system as well as the application of fiber.

A short section describes the issues involved in changing cable television systems to digital transmission. Expect to see changes in coming years.

The final section describes other video applications for fiber optics. They mostly involve requirements for high-quality transmission within buildings or on remote sites, or durable light-weight cables.

<u>Questions to Think About for Chapter 27</u>

1. Analog-to-digital conversion generates lots of extra bits, so it isn't fair to say that a 25-Megabit HDTV frame contains only 2.8 times more information than an NTSC frame digitized to give 9 megabits. It's fairer to compare the number of lines of resolution and the width of the screen. Using those guidelines, how much more information does a 1080-line HDTV image contain than a 525-line NTSC image? Remember that the HDTV image has a 16:9 aspect ratio, while the NTSC image is only 4:3. (*Hint*: calculate the number of picture elements or pixels).

Answer: The HDTV signal has 1920 picture elements across each line (1080x16/9). The NTSC signal has the equivalent of 700 pixels (525×4/3). Calculate the ratio of the pixels and you get the HDTV image contains 5.64 times more than the NTSC image. That difference is double the difference in bits when comparing HDTV to digitized NTSC.

2. The highest resolution possible for digital television is 1080 lines by 1920 pixels, in 60 interlaced frames per second. The lowest is 480 lines by 640 pixels in 24 progressive scans per second (corresponding to a digitized movie). How do the numbers of pixels per second compare? (Note that multiple bits encode each pixel, so this is not the data rate).

Answer: 124 million pixels per second for the high-resolution image vs. 7.4 million for the low-resolution image, a factor of 16.9 difference.

3. Digitizing voice and video both produce data streams with much higher numbers of bits per seconds than the bandwidth in hertz. Compare the ratios of bits per second per hertz for voice and video? What might cause the difference?

Answer: Digitizing a 3-kHz voice signal generates 64 kbit/s, a ratio of 21. Digitizing a 6-MHz video signal generates 270 kbit/s, a ratio of 45. The difference indicates that video engineers chose an algorithm which represented the video signal more accurately than a voice signal. This isn't surprising because the telephone does not try to reproduce voices with high fidelity.

4. A broadcast transmitter in a hybrid fiber-coax system generates output power of 10 dBm (10 mW). Analog receivers require an input power of 5 μW (-23 dBm) for adequate signal-to-noise ratio. If the transmission loss between head end and distribution node is 10 dB (not counting the splitter), and system margin is 10 dB, how many nodes can this transmitter support? How many could you serve by reducing the system margin by 3 dB?

Answer: The loss budget is

Transmitter power: 10 dBm

Cable loss -10 dB

Splitter loss X dB

<u>*System margin -10 dB*</u>

Receiver input -23 dBm

This leaves 13 dB for the splitter. Assuming no excess loss, a 13 dB splitter can divide the signal among 20 nodes. Reducing system margin 3 dB would double the number of nodes to 40.

5. You need narrowcast transmitters for the same system. What power level do they require if system margin, receiver sensitivity, and cable loss are the same?

Answer: The loss budget is

Transmitter power: X dBm

Cable loss -10 dB

<u>*System margin -10 dB*</u>

Receiver input -23 dBm

In this case, transmitter power is -3 dBm.

6. You need to lease capacity on a metro network to transmit one channel of studio-quality HDTV from a studio to a television transmitter. The network operator has four types of transmitters available, which operate at rates to OC-3, OC-12, OC-48, and OC-192. Which one offers the capacity you need without too much excess?

Answer: Studio-quality HDTV requires 1.5 Gbit/s, so you want the 2.5 Gbit/s OC-48 transmitter.

Answers to Quiz for Chapter 27

1. c

2. b

3. d

4. a

5. b

6. a

7. b

8. d

9. c

10. e.

11. a.

12. c.

<u>Supplementary Material</u>

Visit a video studio which uses fiber for high-quality transmission.

Watch for news on digital television, and related ideas such as digital transmission of movies to theaters.

Have students investigate how they receive video signals. Do they subscribe to cable or satellite, or receive over-the-air broadcasts. Cable is most relevant for fiber optics.

Analog television: **http://www.ntsc-tv.com**

Digital television (Advanced Television Systems Committee) : **http://www.atsc.org/**

Chapter 28:

───

Mobile Fiber-Optic Communications

This short chapter is intended to answer questions about special types of fiber-optic communications that may interest students, but are not common. The automotive systems are interesting, and are likely to spread, but they are not yet a main-line fiber-optic application. Unless your students are going into one of those fields, it's a good choice to assign as outside reading rather than covering in class, or to skip if you're pressed for time.

Overview

The common features of the systems described in this chapter is their mobility. They fall into two broad categories. One is cable connections to moving vehicles, such as tethered submarines. The other is cable systems within vehicles, including aircraft, battleships, and automobiles.

Radio control is common for remote control of robotic vehicles, but fiber is better in certain circumstances covered in the second section of this chapter. Fiber-guided missiles on the battlefield are one example; the fiber provides a reliable high-bandwidth link that's immune to interference or jamming. Fiber-guided robots can be used in other hazardous environments on land, particularly in hostile environments. Fiber cables are used as tethers for robotic vehicles undersea, where radio waves cannot penetrate. I mention the most famous example, Robert Ballard's exploration of the sunken wreck of the *Titanic.*

Fibers light weight and immunity to electromagnetic interference make them attractive for use in aircraft and spacecraft. I mention a number of applications in military and civilian aircraft, as well as in the International Space Station. A short section mentions similar—but longer—fiber-optic networks in ships.

The final section covers the introduction of fiber-optic data links into cars. This has taken a long time. Automobiles are a difficult environment for fiber optics because their communication systems must be very cheap, durable, and easy to repair by nonspecialists. However, standards now exist and fibers are being built into some high-end cars, largely for entertainment systems.

Questions to Think About for Chapter 28

1. Military research agencies in Britain and the United States were among the first sponsors of fiber-optic development. Yet military equipment has lagged far behind civilian telecommunications in deploying fiber-optic systems. Why?

Answer: This is a discussion question, with no real "wrong" answers. Important factors include the long lead time for developing military equipment, the need to ruggedize military hardware, and the fact that military vehicles don't need to transmit signals very far, so they don't benefit as much from fiber.

2. One of the first fiber-optic systems deployed by the military was a portable battlefield communications network. The lightweight fiber cables replaced thick copper cables which had proved very vulnerable to damage in handling. Recently wireless systems have replaced the fiber network. Why do you think fiber optics were deployed earlier in battlefield networks than in other systems?

Answer: This is another discussion question. In this case, the first fiber-optic networks were deployed because the old 26-pair copper cables were unwieldy and very vulnerable to damage. Fiber cables were lighter, smaller, much more durable, and much easier to install; no other technology could do the job. The need was not as urgent on aircraft and ships, where existing electronic communication systems worked adequately. Wireless communications later replaced fiber because there was no need to deploy cables.

3. Why can't radio-controlled vehicles be used underwater?

Answer: The radio waves don't penetrate water.

4. A Nimitz-class nuclear aircraft carrier is 1092 feet (333 meters) long and requires a crew of 3300 people to sail. Could you run a gigabit Ethernet link the length of the ship with 62.5/125 μm graded-index fiber at 850 nm? If not, could you use other types of multimode fiber?

Answer: No; the maximum length at 850 nm is 275 m. The alternatives are to use 50/125 μm fiber at 850 nm, or to use either 50/125 or 62.5/125 μm fiber at 1300 nm, all of which could go at least 500 m.

5. A new car is 20 feet (6 m) long. A step-index plastic fiber has bandwidth of 10 MHz/km. Can you use this fiber to run a 25-Mbit/s MOST link the length of the car?

Answer: The nominal bandwidth of such a short length of fiber is over 1 GHz, so it should be no problem.

6. How far can you transmit a 25-Mbit/s NRZ signal through the step-index fiber of Question 5, considering only bandwidth? If attenuation is 200 dB/km, and the link margin is 20 dB, what is the limit imposed by distance, neglecting connector and coupling losses?

Answer: Using the bandwidth relationships in Chapter 20, the allowable pulse spreading for a 25 Mbit/s signal is

$$\Delta t = \frac{0.7}{25 \ Mbit/s} = 28 \ ns$$

This means the signal requires a bandwidth of

$$B(MHz) = \frac{350}{28 \ ns} = 12.5 \ MHz$$

over the length of fiber used. For 10 MHz-km, this means a length of 800 m. A link margin of 20 dB in attenuation corresponds to 100 m of 200 dB/km fiber, so transmission distance is limited by attenuation.

<u>Answers to Quiz for Chapter 28</u>

1. e

2. a

3. c

4. e

5. c

6. b.

7. c.

8. a.

9. e.

10. a.

<u>Supplementary Material</u>

Information on the automotive fiber standard is available at the web sites of the following organizations:

Automotive Multimedia Interface Collaboration: **http://www.ami-c.org**

Byteflight: **http://www.byteflight.com**

Firewire/1394 Trade Association: **http://www.1394ta.com**

Most Cooperation: **http://www.mostcooperation.com**

<h1 style="text-align:center">Chapter 29:</h1>

———

<h1 style="text-align:center">Fiber-Optic Sensors</h1>

I included Chapters 29 and 30 in an effort to cover all of fiber optics. Sensors are a separate field from communications, and if your course is limited to communications, you may want to skip this chapter. However, sensors are important in electronics and control systems, and fiber sensors are emerging as an important new class of devices, so students who plan to work in those areas should read this chapter as an introduction to fiber sensors and their operation.

<h2 style="text-align:center"><u>Overview</u></h2>

I open by describing the crucial difference between fibers for sensing and for communications. Sensing fibers are intended to respond to their environments; communication fibers are not supposed to respond to external influence. If you want to sense pressure, you mount a fiber in a way that makes its loss vary with pressure, while the cable structures used in communication are designed to *isolate* fibers from pressure effects that could influence their transmission.

The oldest family of fiber sensors are probes, which use fibers to deliver and collect light rather than to sense the environment directly. One simple example is a sensor that looks for a part on an assembly line, shown in Figure 29.1. Fibers also can deliver light to other types of optical sensors, called remote optical sensors or optical remote sensors. In both cases, the fibers function as light pipes, not as sensors.

Newer fiber sensors use the fibers themselves as the sensing mechanism. I describe the workings of a thermometer to show the idea of sensing an environmental change, they describe major fiber sensing techniques: intensity modulation, polarization modulation, and phase modulation detected by interferometry.

The final part of the chapter gives some specific examples, including microbending and phase-shift sensors. I pay particular attention to fiber gyroscopes for rotation sensing, and the embedding of fibers in composite materials to make "smart skins" and "smart structures." These two are the fiber sensor applications your students are most likely to encounter.

<h3 style="text-align:center"><u>Questions to Think About for Chapter 29</u></h3>

1. A crack sensor uses a step-index multimode fiber with 100-μm core to detect structural failure. It sets off an alarm when light intensity drops 10 dB. If a crack splits the block of material containing the fiber and causes one side to drop, estimate how far it must drop to set off the alarm. Ignore end reflection effects.

Answer: Light intensity depends on the overlapping areas. For light intensity to drop 10 dB—a factor of 10—only 10% of the fiber cores can overlap. The precise formula for core overlap geometry in this situation is too complex to expect students to work out, but reasonable estimates are 60 to 100 μm.

2. Temperature causes the refractive index of a fiber to increase by 0.001% per degree. Two arms of an interferometric fiber sensor each contain 1 cm of fiber. One arm is exposed to the environment, the other is kept at a constant temperature. If you use 1-μm light, how much temperature change is needed to produce a 180° phase shift?

101

Answer: The change in refractive index changes the effective length of a 1-cm length of fiber by 10^{-5} per degree, so light in the warmer arm is retarded by 0.1 µm per degree. At a wavelength of 1 µm, a 180° phase shift is 0.5 µm, which would be produced by a 5.0° temperature increase. [NOTE: INITIAL VERSION OF THIS ANSWER ASSUMED A ROUND TRIP THROUGH THE FIBER FOR TOTAL TRANSMISSION OF 2 CM, BUT THAT WAS NOT STATED. THE CONFIGURATION SHOWN IN THE BOOK WAS A SINGLE-PASS.]

3. There are two different ways you could increase the sensitivity of the interferometric sensor in question 2 so the sensor measures a 1° temperature change with a 180° phase shift, without changing the glass used in the fiber. What are they?

Answer: Increase the length of the fiber to 5 µm, so the light in the warmer arm fall further behind, or decrease the wavelength to 0.4 µm. so the 0.2-µm phase change per degree would equal a 180° phase shift.

4. A fiber-optic gyroscope includes a one-meter loop of fiber and a laser light source emitting at 1 µm. How much rotation does it take to cause a 180° phase shift between the two counter-propagating beams?

Answer: One wave shifts 90° in one direction while the other wave shifts 90° in the other direction to produce a 180° phase shift. This means the ring of fiber must shift one-quarter of a 1-µm wavelength around a 1-m loop which is 10^9 wavelengths long. This rotation is $0.25 \times 10^{-9} \times 360$ degrees, or 9×10^{-8} degree.

5. Recall that light travels roughly 3×10^8 m/s. What rotation rate does the 180° phase shift in question 4 correspond to if it's detected over the time circles through the fiber once?

Answer: Light travels 1 meter in 3.3×10^{-9} second. The ring has to rotate 9×10^{-8} degree in that interval, at a rate of 27 degrees a second. That sounds fast, but remember that's what it takes to detect a 180° phase shift in an interval of 3.3 nanoseconds. Actual fiber gyros use longer fibers in on a spool and monitor the phase shift continually to keep track of the cumulative rotation.

<u>Answers to Quiz for Chapter 29</u>

1. a

2. b

3. b

4. d

5. c

6. c

7. b

8. c

9. b

10. e

<u>Supplementary material</u>

The simplest fiber sensor to demonstrate without elaborate equipment is microbending. Try it with multimode fiber sandwiched between sheets of the coarsest sandpaper you can find, but be sure to test it beforehand because microbending losses vary widely.

Chapter 30:

Imaging and Illuminating Fiber Optics

Optical fibers were first used in bundles to transmit images and illumination. Imaging and illumination remain important applications of fiber optics, but they are of limited interest for students specializing in communications. You should cover this chapter if your students are studying instrumentation, imaging, optics, or electronics. If you are not familiar with this use of fiber optics, you should read the chapter yourself so you can answer student questions.

Overview

So far, I have treated optical fibers individually, as if the only place they contacted other fibers was at their ends where light was coupled between them. Here, I explain how fibers are arranged in bundles to transmit images and light for illumination. Coherent bundles are made with fibers aligned on their two ends so they can carry images; they may be flexible (with fibers loose in the middle), or rigid, with fibers fused together. Fibers may be arranged randomly for illumination, or they may be bundled in groups to form patterns, such as in illuminated signs.

You can think of each fiber in an imaging bundle as carrying a single pixel of light, as shown in Figure 30.3, which means the finer the fibers, the clearer the image. In practice, the fibers are small enough that you don't notice individual fibers unless you examine the bundle with a magnifier. The optics of imaging bundles differ subtly from those of single fibers, primarily because the simplifications possible for single fibers do not work with bundles. This can produce some interesting effects, including magnification or shrinking of images.

Individual fibers also can be used as optical elements. One interesting example is the use of short sections of graded-index fibers as microlenses. Thick single fibers are used to direct laser beams to targets both in industrial materials-working, and in precision micro-surgery.

I also describe several specific uses of both rigid and flexible fibers. Short rigid bundles are used as image inverters, faceplates, and other optical elements in some military systems; their advantage is efficient transfer of light over a short working distance. Flexible bundles are used in medical examination and treatment, and for industrial inspection.

Fiber-optic light pipes are used for illumination. Single large-core fibers can deliver high powers for surgery or materials-working. Bundles of fibers can be arranged with their ends in a pattern to form a sign, as shown schematically in Figure 30.9.

Questions to Think About for Chapter 30

1. Standard laser printers used to have resolution of 300 dots per inch. Could you spot those dots with a bundle of 10-μm fibers? (Assume that resolution in line pairs is equivalent to dots per inch.)

Answer: Convert the bundle resolution of 50 line pairs per millimeter to inches. One inch is 25.4 inches, so 50 line pairs per millimeter is 1270 lines per inch. You should have no problem.

2. Using the criteria that the finest possible resolution corresponds to half a line pair per fiber core diameter, what is the largest core fiber that could resolve 300 line pairs per inch?

Answer: 300 line pairs per inch is roughly 12 line pairs per millimeter. The largest core diameter able to resolve that is 1/24 mm or 42 μm.

3. A typical packing fraction for bundled fiber is 90%. Assuming that the fiber cores are 10 μm and you can neglect space between fibers (not a good assumptions, but it makes the math manageable), what is the outer diameter of each fiber and how thick is the cladding?

Answer: Calculate from the ratio of the areas of the core and the clad fiber:

$$\frac{\pi(10)^2}{\pi(\text{cladding})^2} = 0.9$$

Solving for outer diameter of the clad fiber gives 10.54 μm. That means thickness of the cladding is only 0.27 μm in this simple approximation.

4. Given the same assumptions about neglecting the spacing between fibers, what would be the packing fraction for a bundle assembled from 100/140 step-index multimode fibers?

Answer: The packing fraction is

$$Packing = \frac{core\ area}{total\ area} = \frac{\pi(100)^2}{\pi(140)^2} = 0.51$$

or 51%. Not very good.

5. You are trying to deliver a 50-watt laser beam through a 3-meter length of fiber with attenuation of 10 dB/km. How much power is lost in the fiber?

Answer: Attenuation in 3 m of 10 dB/km fiber is 0.03 dB. Use the formula for attenuation in dB:

$$loss(dB) = -10\log_{10}\left(\frac{P_{out}}{P_{in}}\right)$$

Plug the numbers in, take the exponent of both sides to remove the log factor, and turn the equation around to solve for P_{out} gives you $P_{out} = 10^{-0.003} \times 50\ W = 49.658$, meaning the loss is only 0.342 W.

6. If you put a fiber-optic image inverter flat on a printed sheet of paper, you can read the inverted letters through the taper. You're seeing reflected light. How did it reach the paper?

Answer: Down through the fibers from ambient light in the room.

<u>ANSWERS TO QUIZ FOR Chapter 30</u>

1. b

2. d

3. c

4. c

5. d

6. b

7. d

8. a

9. d

10. b

<u>Supplementary Material</u>

Fiber-optic Christmas ornaments, some of which attached to small bulbs, are great to pass around for students to examine—and cheap enough that you don't have to worry much about them "disappearing".

If you have some plastic fibers, you can easily make your own fiber-optic sign; illuminate it from behind to show the effect if you can make your room dark enough.

Short rigid bundle components such as image inverters or imaging guides are fascinating, and rejects that don't meet stringent military requirements are sometimes available. However, you should watch them carefully; they are so fascinating that students may be tempted to take them home.

General Resources

The list of resources that follows is a starting point for further inquiries into fiber-optic technology. If you teach fiber optics, you should subscribe to the major trade magazines in the field to monitor development of the field. Major print magazines all have web sites; if you don't find them immediately, just search for them. I have checked all the web sites listed as of March 2005. Some corporate and academic web sites include tutorial articles on their own work, which are best uncovered by web searches. The organizations listed are all American with one notable exception, the Institution of Electrical Engineers, which publishes one of the best research journals in the field, *Electronics Letters*, which unfortunately is quite expensive.

No single professional society serves the bulk of the fiber optics community. The Fiber Optic Association claims about 7000 members, and is largely directed at technicians. The Optical Society of America and SPIE are optics-oriented groups, and both pay particular attention to optics-related education. IEEE is oriented toward electronics, but includes fiber-optic communications and other electro-optical technologies in its domain. The Laser Institute of America has strong programs in safety issues related to lasers.

Printed trade magazines specializing in fiber optics:

Fiberoptic Technology, 100 Enterprise Drive, Suite 600, Box 912, Rockaway, NJ 07866-0912, (973) 920-7000; **http://www.fiberoptictechnology.net/**

Lightwave, 98 Spit Brook Rd., Nashua, NH 03062, 603-891-0123,

http://www.light-wave.com

Web-Only publications specializing in fiber optics:

Fiberoptics Online (**http://www.fiberopticsonline.com**)

Fibers.org (**http://www.fibers.org**)

Light Reading (**http://www.lightreading.com**)

Trade magazines covering related areas:

Electronic Engineering Times, **http://www.eetimes.com**

Laser Focus World, 98 Spit Brook Rd., Nashua, NH 03062, 603-891-0123, **http://www.laserfocusworld.com**

Network Magazine **http://www.networkmagazine.com/**

Photonics Spectra, Berkshire Common, PO Box 4949, Pittsfield, MA 01202, 413-499-0514, **http://www.Photonics.com**

Telephony, One IBM Plaza, Suite 2300, Chicago, IL 60611, 312-595-1080, **http://telephonyonline.com/**

Professional Societies and Associations

Fiber Optic Association, **http://www.thefoa.org**

Institute of Electrical and Electronics Engineers (IEEE), 445 Hoes Lane, P.O. Box 1331, Piscataway, NJ 08855-1331, 732-562-6820, **http://www.ieee.org**

Institution of Electrical Engineers (IEE), Savoy Place, London, WC2R 0BL, UK, tel 011-44-171-240 1871, Fax: +011-44-171 240 7735, **http://www.iee.org**

Laser Institute of America, 12424 Research Parkway, Suite 125, Orlando, FL 32826, 407-380-1553, **http://www.LaserInstitute.org**

Optical Society of America, 2010 Massachusetts Ave. NW, Washington, DC 20036 202-223-8130, **http://www.osa.org**

New England Fiberoptics Council, **http://www.nefc.com**

SPIE (The International Society for Optical Engineering), PO Box 10, Bellingham, WA 98227-0010, 360-676-3290, **http://www.spie.org**

Technical journals

Electronics Letters (published by Institution of Electrical Engineers)

IEEE Communications Magazine

IEEE Journal on Selected Areas in Communications

IEEE Photonics Technology Letters

Journal of Lightwave Technology (published by IEEE and OSA)

Books

General Fiber Optics

Vivek Alwayn, *Optical Network Design and Implementation* (Cisco Press, 2004) (Telecommunication systems using fibers)

M. Bass ed., *OSA Handbook of Optics Vol. 4 Fiber Optics and Nonlinear Optics* (McGraw Hill, New York, 2001) (massive handbook)

John A. Buck, *Fundamentals of Optical Fibers* (Wiley Interscience, New York, 1995) (textbook on fibers and their properties)

C. David Chaffee, *Building the Global Fiber Optics Superhighway* (Kluwer Academic, 2001) (recent history of global telecommunications)

Bob Chomycz, *Fiber Optic Installer's Field Manual* (McGraw Hill, New York, 2000) (for the practicing technician)

Dennis Derickson, *Fiber Optic Test and Measurement* (Prentice Hall, Upper Saddle River, NJ, 1998) (excellent coverage)

David R. Goff, *Fiber Optic Reference Guide: A Practical Guide to the Technology, 2nd edition* (Focal Press, 1999) (concise reference guide)

Jim Hayes, *Fiber Optics Technician's Manual* (Delmar Publishers, Albany, 1996) (for the practicing technician)

Jeff Hecht, *City of Light: The Story of Fiber Optics* (Oxford University Press, New York, 1999, updated paperback edition 2004) (history, easy reading)

Gerd Keiser, 3rd ed, *Optical Fiber Communications* (McGraw Hill, New York, 2000) (recommended as the next step up from this book)

Rajiv Ramaswami and Kumar N. Sivarajan, *Optical Networks, A Practical Perspective* (Morgan Kaufmann, 2002) (fiber-optic telecommunications systems)

Telecommunications in General

John C. Bellamy, *Digital Telephony 3rd ed.,* (Wiley, 2000)

Walter Ciciora, James Farmer, and David Large, *Modern Cable Television Technology: Video, Voice, and Data Communications* (Morgan Kaufmann, San Francisco, 1999) (very comprehensive)

Alan Freedman, *The Computer Glossary* (American Management Association, New York, 2001) (Includes many telecommunications terms)

Roger L. Freeman, *Fundamentals of Telecommunications* (Wiley-Interscience, New York, 1999 (oriented toward systems, at the engineering level)

Gary M. Miller, *Modern Electronic Communications* 6th ed (Prentice Hall, Upper Saddle River, NJ, 1999 (good nuts and bolts introduction)

Laszlo Solymar *Getting the Message: A History of Communications* (Oxford University Press, 1999) (an entertaining and enjoyable history)

Optics, Photonics, and Lasers

Govind P. Agrawal ed., *Semiconductor Lasers: Past, Present, and Future,* (AIP Press, Woodbury, NY, 1995) (good overview)

Michael Bass, ed., *Handbook of Optics: Vol. 1: Fundamentals, Techniques & Design* and *Vol. 2: Devices, Measurements and Properties* (McGraw Hill, 1995) (massive, comprehensive handbooks)

J. Warren Blacker and Peter Schaeffer, *Optics, An Introduction for Technicians and Technologists* (Prentice Hall, 2000), (Introductory)

Eugene Hecht *Optics 4th ed* (Addison Wesley, Reading, MA, 2001) (the standard upper-level undergraduate optics text; no relation)

Jeff Hecht, *Understanding Lasers* (IEEE Press, Piscataway, NJ 1994) (introductory level, but a bit dated on semiconductor lasers)

Francis A. Jenkins and Harvey E. White *Fundamentals of Optics 4th ed* (McGraw Hill, New York, 1976) (very old, but a good introduction to classical optics)

B. A. E. Saleh and M. C. Teich *Fundamentals of Photonics*, (Wiley-Interscience, New York, 1991) (wide-ranging upper-level textbook)

Amnon Yariv *Optical Electronics in Modern Communications 5th ed* (Oxford University Press, New York, 1997) (widely used but very advanced textbook)

Additional Acronyms (supplement to Glossary)

ATV Advanced Television (U.S. digital television standard including high definition)

BER Bit error rate

BOC see RBOC.

CO Central office, telephone switching office.

CPE Customer premises equipment, your phone, fax, modem etc.

CWDM Coarse wavelength-division multiplexing

DSx A signal rate in the North American Telephone Hierarchy, corresponding to a T carrier; e.g., DS1 is the rate of signals transmitted by a T1 line

FDDI Fiber distributed data interface

G-Lite A type of DSL

IDSL ISDN repackaged as a DSL service

IEC Interexchange carrier, a long distance-company

ILEC Incumbent local exchange carrier—the dominant local telephone company

ISDN Integrated Services Digital Network, a digital service over phone lines

ISP Internet service provider, company offering Internet service, may be local or national.

ITU International Telecommunications Union

IXC Interexchange carrier, a long-distance phone carrier.

LAN Local area network

LEC Local exchange carrier, the established local phone company.

NRZ No return to zero (digital signal coding)

PCS Personal communication service, a digital mobile phone service, functionally similar to cellular phones, which use analog transmission.

PCS fiber Plastic clad silica fiber

QOS Quality of service.

RBOC Regional Bell operating company, established local phone company that was split from AT&T.

RHC Regional holding company, established local phone company, usually one that was split from AT&T.

RZ Return to zero (digital signal coding)

SNA Systems network architecture.